AF356135

Managing the Product Quality of Vegetable Crops under Abiotic Stress

Managing the Product Quality of Vegetable Crops under Abiotic Stress

Editor

Lilian Schmidt

MDPI • Basel • Beijing • Wuhan • Barcelona • Belgrade • Manchester • Tokyo • Cluj • Tianjin

Editor
Lilian Schmidt
Geisenheim University
Germany

Editorial Office
MDPI
St. Alban-Anlage 66
4052 Basel, Switzerland

This is a reprint of articles from the Special Issue published online in the open access journal *Horticulturae* (ISSN 2311-7524) (available at: https://www.mdpi.com/journal/horticulturae/special_issues/Vegetable_Crops_Horticultural).

For citation purposes, cite each article independently as indicated on the article page online and as indicated below:

LastName, A.A.; LastName, B.B.; LastName, C.C. Article Title. *Journal Name* **Year**, *Volume Number*, Page Range.

ISBN 978-3-0365-3098-7 (Hbk)
ISBN 978-3-0365-3099-4 (PDF)

Cover image courtesy of Lilian Schmidt

© 2022 by the authors. Articles in this book are Open Access and distributed under the Creative Commons Attribution (CC BY) license, which allows users to download, copy and build upon published articles, as long as the author and publisher are properly credited, which ensures maximum dissemination and a wider impact of our publications.

The book as a whole is distributed by MDPI under the terms and conditions of the Creative Commons license CC BY-NC-ND.

Contents

 *horticulturae*

MDPI

Editorial

Managing the Product Quality of Vegetable Crops under Abiotic Stress

Lilian Schmidt [1,2]

[1] Professorship for Vegetable Crops, Hochschule Geisenheim University, Von-Lade-Strasse 1, 65366 Geisenheim, Germany; Lilian.Schmidt@hs-gm.de

[2] Professorship for Soil Science and Plant Nutrition, Hochschule Geisenheim University, Von-Lade-Strasse 1, 65366 Geisenheim, Germany

Citation: Schmidt, L. Managing the Product Quality of Vegetable Crops under Abiotic Stress. *Horticulturae* **2022**, *8*, 25. https://doi.org/10.3390/horticulturae8010025

Received: 1 December 2021
Accepted: 14 December 2021
Published: 26 December 2021

Publisher's Note: MDPI stays neutral with regard to jurisdictional claims in published maps and institutional affiliations.

Copyright: © 2021 by the author. Licensee MDPI, Basel, Switzerland. This article is an open access article distributed under the terms and conditions of the Creative Commons Attribution (CC BY) license (https://creativecommons.org/licenses/by/4.0/).

1. Introduction

Plants, as sessile organisms, are continuously exposed to varying environmental conditions and often face abiotic and biotic threats. Biotic stressors include pathogens such as fungi, bacteria or herbivores that can damage plant tissues locally and systemically. In addition to this, abiotic stress, such as sub- and supra-optimal temperatures, the availability of water, light conditions and nutrient supplies, may severely impact the performance of plants. This includes the effects on photosynthetic processes, growth and developmental alterations and modifications in metabolic pathways.

In consequence, plants have developed a wide range of strategies to cope with the challenging conditions, mainly the alteration of growth processes and the investment in defensive compounds [1]. These two strategies often represent a dilemma for plants as they require different resource allocation patterns.

When cultivating plants for use in human nutrition, farmers are faced with a similar dilemma: Besides a high crop yield, there is an increasing demand for products that are rich in health-promoting substances. These beneficial compounds are mostly secondary plant metabolites that are synthesized as a consequence of abiotic or biotic stresses. This is especially true for vegetable crops that are rich in secondary compounds such as polyphenols or carotenoids. In addition, vegetables can contain substances with undesired properties, e.g., oxalates and nitrate. It is thus of interest to cultivate vegetable crops with maximum contents of beneficial compounds and minimal contents of unfavorable compounds, as well as high yields.

Significant research is focusing on pre- and post-harvest procedures that can be used to achieve these goals. Abiotic stress is often applied by employing certain cultivation practices and storage conditions in order to maintain and potentially improve the internal and external properties of vegetable crops.

2. Special Issue Overview

This Special Issue compiles 11 manuscripts that deal with the effects of abiotic stress on the modulation of vegetable quality. It comprises 10 research articles and 1 review. The majority of the contributions deal with pre-harvest conditions and investigate whether cultivation practices can modulate the yield and nutritional value of vegetable crops and herbs under abiotic stresses. Among these publications, several deal with nutrient supply [2–5], salinity [6,7], water supply [2,3,8] or lighting conditions [6]. Two manuscripts describe either the effects of applying chemical substances such as gibberellic acid [9] or seaweed extract [10] on the yield and quality of vegetable crops. Grafting as a method to mitigate the impacts of abiotic stress in vegetables was elucidated by a research paper by Ellenberger et al. [3] and in a review by Devi et al. [11]. Elevated CO_2 levels were either used as pre-harvest conditions in combination with different nitrogen forms [5] or as post-harvest treatment combined with different methods of packaging [12].

In general, a large quantity of the contributions in this Special Issue present data on fruit vegetables such as watermelons, bitter gourd and tomatoes [2–4,9,11], with the latter being the focus of three publications. These crops were mostly grown in greenhouses [2–4], but some were grown in an open field [9] or under rain-out shelters [8]. Leafy vegetables and herbs—such as arugula [5], purslane [6] and basil [7]—were used in three experiments and were cultivated in growth cabinets [5,6] or in greenhouses [7]. The bulbs of onions were harvested from plants in a field trial [10], and the asparagus spears for a storage experiment were obtained from local farmers [12].

Altogether, a wide range of species, cultivation conditions and management practices (pre- and post-harvest) were used for elucidating the effect of abiotic stress on the quality of vegetable crops.

2.1. Impacts of Reduced Water Supply

The availability of water for the irrigation of vegetable crops will become of high relevance in the future; while heavy rainfall is predicted to occur more frequently, the opposite will be the case as well, with more and longer drought events as compared to the present conditions. One way to overcome these challenges is the use of vegetable varieties that are able to cope with these extreme conditions. Besides the breeding activities that aim at creating more resistant varieties, the use of traditional varieties that may be better adapted to specific situations is being increasingly suggested. In their study, Sica et al. [8] tested the local bean accessions of Veneto, Italy, concerning the yield potential and nutritional properties of the seeds as consequences of drought exposure during different growth stages. Interestingly, exposure to drought had no impact on the yield of dry seeds until the flowering phase. In general, most impacts on the nutraceutical properties of the bean seeds could be explained by variety, and rather little can be assigned to the drought treatments. One bean accession turned out to be most promising for cultivation in areas with increased risk of drought events.

2.2. Effects of Alterations in Nutrient Supply in Combination with Other Abiotic Stresses

Mineral nutrients are essential for plants. When applied in concentrations or forms that are not optimal for the crops, this may have consequences for the performance of plants and their products. Three contributions to this Special Issue deal with yield and quality aspects of tomatoes under different nutritional conditions that are regarded as "stressful" for the plants. Specifically, Ellenberger et al. [3] addressed the question of how grafting two commercial tomato varieties may affect the yield and fruit quality when exposed to a combination of nutrient deficiency and drought stress. They used the commercial rootstock 'Beaufort', two novel hybrid rootstocks of *S. pennellii* × *S. lycopersicum*, which are expected to be drought tolerant, and self-grafted plants. When exposed to a 50% reduction in fertigation, the 'Lyterno' variety grafted onto the novel rootstocks had higher fruit yields as compared to the plants grafted either on the commercial cultivar or the self-grafted plants. This was not observed for the tomato cultivar 'Tastery', which, in general, showed weak responses to the imposed stress. For all the combinations of grafting and fertigation supply, little effects in terms of fruit quality were assessed. The use of appropriate scion–rootstock combinations may thus provide potential for optimizing the stress resilience, the yield and the quality of tomatoes.

Moreover, de Luca et al. [2] combined potassium (K) supply and water shortage treatments for tomatoes, using a commercial cultivar and a transgenic line that may accumulate large quantities of K. The study aimed at testing whether different K supplies can alleviate the impact of periods of dehydration. Although it was obvious that the larger supply of K was able to increase the shoot growth and fruit yield under water stress, this did not result in yields comparable to those obtained for well-watered plants. Moreover, the transgenic tomato line with increased capacities to accumulate K turned out not to be beneficial under water stress in terms of fruit yield.

In another study on tomatoes, Schmidt and Zinkernagel [4] investigated the effects of fertilization with 50% less nitrogen (N) than recommended on the quality of three different varieties of cocktail tomatoes immediately after harvest and after eight days of storage at low temperatures. It was shown that the reduction in N had little effects on the yield, external quality and on the concentrations of valuable compounds in the tomato fruits. In addition, there were no impacts on the taste, as observed by a trained sensory panel, suggesting that the reduced N supply was not detrimental for the three varieties of tomato, at least under the experimental conditions.

One manuscript describes the effects of elevated atmospheric CO_2 levels and N forms on the physiological processes, crop yield and nutritional quality of the leafy vegetable arugula [5]. Plants of two arugula varieties were grown in climate cabinets with either 400 or 800 ppm CO_2 and received either pure nitrate or ammonium-dominated nitrogen fertilizer. Replacing 75% of nitrate-N with ammonium-N had little impact on photosynthetic CO_2 assimilation, the nutritional quality (e.g., content of anti-oxidative compounds) or was disadvantageous for crop yield, regardless of the CO_2 levels. Thus, ammonium-dominated N provides no benefits for arugula production under the elevated CO_2 levels that are proposed for the future.

2.3. Impact of Salinity

Salinity is characterized by increased electrical conductivity due to an accumulation of mineral elements. Specifically, the enrichment with sodium (Na) and chloride (Cl) ions in the soil imposes a worldwide threat to plant production. However, as a mild stressor, saline conditions can be used to alter the quality of horticultural products.

This was conducted by Corrado et al. [7] by supplying sweet basil plants in hydroponics with three nutrient solutions with differing NaCl concentrations. Even low concentrations of NaCl (20 mM) resulted in reduced biomass development, while having positive effects on some nutritional quality parameters, such as antioxidant capacity and total polyphenols.

In a study by Giménez et al. [6], much higher salinity levels (80 mM of NaCl) were combined with different light conditions (fluorescent lamps, red–blue LED, red–blue–deep red LED). Purslane coped well with the salinity level under the two LED lighting regimes. The nutritional quality of this microgreen was improved as anti-nutritive compounds were lower when grown with LEDs and supply of NaCl. The study showed that it is possible to achieve high yields when growing purslane microgreens under saline conditions and an appropriate choice of light sources, although this was at the expense of some valuable plant pigments (chlorophylls, carotenoids, flavonoids).

2.4. Effects of Application of Different Substances

Stressful conditions can also be imposed by applying certain chemicals to plants. In particular, substances that mimic the action of plant hormones are the focus of attention. Gibberellic acid is known to regulate both vegetative and reproductive growth processes in plants. It thereby affects the quality of the product by enhancing its antioxidant status. Both the alteration of plant growth and developmental processes and the impact on the nutritional quality are very relevant for the tropical vegetable crop *Momordica charantia* (bitter gourd). Abbas et al. [9] conducted field trials with two varieties of bitter gourd exposed to five different concentrations of a commercial gibberellic acid product. They found that the exogenous application of the plant growth regulator at high concentrations enhances growth (e.g., increased length of petioles and internodes) and yield (increased fruit numbers, increased fruit weight) while decreasing the activity of several antioxidant enzymes at the same time.

In another contribution, Abbas et al. [10] investigated the effects of foliar spraying with seaweed extract on the productivity and quality of four onion cultivars. The bio-stimulant was applied at four concentrations, with the lowest concentration (0.5%) resulting in an increased yield, increased contents of some mineral nutrients and increased soluble solids of the bulbs. The highest concentration of the seaweed extract (3%) increased the concentrations of ascorbic acid in the bulbs as compared to those without the application of seaweed extract.

2.5. Grafting as a Method to Reduce the Impact of Abiotic Stress

Grafting is a commonly used method in overcoming abiotic and biotic stresses during the cultivation of fruit vegetable crops. The advantage of this cultivation technique is the combination of potential beneficial properties of two or more accessions in one plant.

Devi et al. [11] contributed a review dealing with the effects of grafting on the fruit maturity and quality of watermelons. The publication summarizes the results of research activities on the impact of grafting watermelon scions on different rootstocks, with a focus on the formation of hollow hearts and hard seeds, flesh firmness, total soluble solid content, pH, titratable acidity, lycopene content and concentration of the non-essential amino acid citrulline, which can be of relevance for the cardiovascular system.

2.6. Modulation by Storage Conditions

The impacts of post-harvest conditions on the quality of vegetables are represented as well. Wang et al. [12] combined elevated CO_2 pre-treatments and different packaging methods in order to assess the effects on different quality characteristics of green asparagus in cold storage. The authors showed that continuous application of elevated CO_2 reduced the decay of the asparagus spears to the greatest extent by inhibiting the microbial development. In general, elevated CO_2 application, either continuously or as a three-day pre-treatment, did not only reduce the microbial load and the respiration rate but also improved the sensory and nutritional quality of the green asparagus. In combination with modified atmosphere packaging, the pre-treatment with elevated CO_2 was able to preserve the visual quality and the fresh weight of the asparagus and reduce the development of an off-odor.

Funding: This research received no external funding.

Acknowledgments: I gratefully acknowledge all the authors that participated in this Special Issue.

Conflicts of Interest: The author declares no conflict of interest.

References

1. Herms, D.A.; Mattson, W.J. The Dilemma of Plants: To Grow or Defend. *Q. Rev. Biol.* **1992**, *67*, 283–335. [CrossRef]
2. De Luca, A.; Corell, M.; Chivet, M.; Parrado, M.A.; Pardo, J.M.; Leidi, E.O. Reassessing the Role of Potassium in Tomato Grown with Water Shortages. *Horticulturae* **2021**, *7*, 20. [CrossRef]
3. Ellenberger, J.; Bulut, A.; Blömeke, P.; Röhlen-Schmittgen, S. Novel *S. pennellii* × *S. lycopersicum* Hybrid Rootstocks for Tomato Production with Reduced Water and Nutrient Supply. *Horticulturae* **2021**, *7*, 355. [CrossRef]
4. Schmidt, L.; Zinkernagel, J. Opportunities of Reduced Nitrogen Supply for Productivity, Taste, Valuable Compounds and Storage Life of Cocktail Tomato. *Horticulturae* **2021**, *7*, 48. [CrossRef]
5. Schmidt, L.; Zinkernagel, J. For a Better Understanding of the Effect of N Form on Growth and Chemical Composition of C_3 Vascular Plants under Elevated CO_2—A Case Study with the Leafy Vegetable *Eruca sativa*. *Horticulturae* **2021**, *7*, 251. [CrossRef]
6. Giménez, A.; Del Martínez-Ballesta, M.C.; Egea-Gilabert, C.; Gómez, P.A.; Artés-Hernández, F.; Pennisi, G.; Orsini, F.; Crepaldi, A.; Fernández, J.A. Combined Effect of Salinity and LED Lights on the Yield and Quality of Purslane (*Portulaca oleracea* L.) Microgreens. *Horticulturae* **2021**, *7*, 180. [CrossRef]
7. Corrado, G.; Vitaglione, P.; Chiaiese, P.; Rouphael, Y. Unraveling the Modulation of Controlled Salinity Stress on Morphometric Traits, Mineral Profile, and Bioactive Metabolome Equilibrium in Hydroponic Basil. *Horticulturae* **2021**, *7*, 273. [CrossRef]
8. Sica, P.; Galvao, A.; Scariolo, F.; Maucieri, C.; Nicoletto, C.; Pilon, C.; Sambo, P.; Barcaccia, G.; Borin, M.; Cabrera, M.; et al. Effects of Drought on Yield and Nutraceutical Properties of Beans (*Phaseolus* spp.) Traditionally Cultivated in Veneto, Italy. *Horticulturae* **2021**, *7*, 17. [CrossRef]
9. Abbas, M.; Imran, F.; Iqbal Khan, R.; Zafar-ul-Hye, M.; Rafique, T.; Jameel Khan, M.J.; Taban, S.; Danish, S.; Datta, R. Gibberellic Acid Induced Changes on Growth, Yield, Superoxide Dismutase, Catalase and Peroxidase in Fruits of Bitter Gourd (*Momordica charantia* L.). *Horticulturae* **2020**, *6*, 72. [CrossRef]
10. Abbas, M.; Anwar, J.; Zafar-ul-Hye, M.; Iqbal Khan, R.; Saleem, M.; Rahi, A.A.; Danish, S.; Datta, R. Effect of Seaweed Extract on Productivity and Quality Attributes of Four Onion Cultivars. *Horticulturae* **2020**, *6*, 28. [CrossRef]

11. Devi, P.; Perkins-Veazie, P.; Miles, C. Impact of Grafting on Watermelon Fruit Maturity and Quality. *Horticulturae* **2020**, *6*, 97. [CrossRef]
12. Wang, L.-X.; Choi, I.-L.; Kang, H.-M. Correlations among Quality Characteristics of Green Asparagus Affected by the Application Methods of Elevated CO_2 Combined with MA Packaging. *Horticulturae* **2020**, *6*, 103. [CrossRef]

horticulturae

Article

Reassessing the Role of Potassium in Tomato Grown with Water Shortages

Anna De Luca [1,2], Mireia Corell [3], Mathilde Chivet [4], M. Angeles Parrado [1], José M. Pardo [2] and Eduardo O. Leidi [1,*]

[1] Department of Plant Biotechnology, IRNAS-CSIC, Avenida Reina Mercedes 10, 41012 Sevilla, Spain; anna.luca@ibvf.csic.es (A.D.L.); maparrado@irnase.csic.es (M.A.P.)
[2] IBVF-CSIC, Avenida Americo Vespucio 49, 41092 Sevilla, Spain; jose.pardo@csic.es
[3] Departamento de Ciencias Agroforestales (ETSIA), Universidad de Sevilla, Ctra. Utrera Km. 1, 41013 Sevilla, Spain; mcorell@us.es
[4] Grenoble Institut Neurosciences (GIN), Université Grenoble Alpes, Inserm, U1216, 38000 Grenoble, France; mathilde.chivet@gmail.com
* Correspondence: eo.leidi@csic.es

Abstract: Potassium (K) is closely related to plant water uptake and use and affects key processes in assimilation and growth. The aim of this work was to find out to what extent K supply and enhanced compartmentation might improve water use and productivity when tomato plants suffered from periods of water stress. Yield, water traits, gas exchange, photosynthetic rate and biomass partition were determined. When plants suffered dehydration, increasing K supply was associated with reduction in stomatal conductance and increased water contents, but failed to protect photosynthetic rate. Potassium supplements increased shoot growth, fruit setting and yield under water stress. However, increasing the K supply could not counteract the great yield reduction under drought. A transgenic tomato line with enhanced K uptake into vacuoles and able to reach higher plant K contents, still showed poor yield performance under water stress and had lower K use efficiency than the control plants. With unlimited water supply (hydroponics), plants grown in low-K showed greater root hydraulic conductivity than at higher K availability and stomatal conductance was not associated with leaf K concentration. In conclusion, increasing K supply and tissue content improved some physiological features related to drought tolerance but did not overcome yield restrictions imposed by water stress.

Keywords: *Solanum lycopersicum*; drought potassium; vacuolar transporter

Citation: De Luca, A.; Corell, M.; Chivet, M.; Parrado, M.A.; Pardo, J.M.; Leidi, E.O. Reassessing the Role of Potassium in Tomato Grown with Water Shortages. *Horticulturae* **2021**, *7*, 20. https://doi.org/10.3390/horticulturae7020020

Received: 23 October 2020
Accepted: 25 January 2021
Published: 29 January 2021

Publisher's Note: MDPI stays neutral with regard to jurisdictional claims in published maps and institutional affiliations.

Copyright: © 2021 by the authors. Licensee MDPI, Basel, Switzerland. This article is an open access article distributed under the terms and conditions of the Creative Commons Attribution (CC BY) license (https://creativecommons.org/licenses/by/4.0/).

1. Introduction

Potassium (K) is the most abundant cation and plays important roles in plant growth and development where it contributes to charge balance, osmotic adjustment and enzyme catalysis [1]. Stored in vacuoles, K contributes to generating the osmotic pressure required for turgor and expansive growth. Potassium contributes to water uptake and the generation of turgor pressure in the stomatal guard cells and the phloem [1]. At low K availability, plants are more susceptible to wilting in drying soils and a mild K deficiency affects photosynthesis by reducing stomatal conductance and photosynthetic biochemical reactions [2]. High K availability might counteract the inhibitory effect of drought on CO_2 assimilation by protecting photosynthetic capacity [3,4].

Plant K uptake is affected by available water as low soil water content reduces mass flow and diffusion of the nutrients to the rhizosphere, and simultaneously decreases root growth [5]. Potassium transport into aerial parts does not rely only on leaf transpiration, and other factors such as root pressure contribute to long distance transport [6]. However, greater water flow by active transpiration increases K flux to the shoot and K uptake by the roots when it is available [7]. In doing so, symplastic K transport depends on membrane transport proteins such as high-affinity KUP/HAK transporters and K-selective channels, which facilitate cation movement between plant cells and its final delivery into the xylem [8]. During drought, K fluxes into shoots become reduced by less transpired water (lower xylem water flow) and reduced expression of K-loading channels into the xylem [7].

Tomato requires high K supplements for high yields and optimum fruit quality [9] and low K affects flower development and fruit setting efficiency [10]. When K is scarce, tomato plants reduce sink activity (stems and fruits) prior to having any effect on photosynthesis [11,12]. Considering the great importance of K supply and water availability for fruit yield and quality, we aimed at studying how K availability and compartmentation into vacuoles might affect plant water use and productivity under water stress. The overexpression of a vacuolar K,Na/H antiporter (NHX1) contributes to salt-tolerance in tomato by increasing root K uptake and its storage plant vacuoles [8,13]. By including a transgenic line able to take up more of this nutrient, we tried to assess to what extent a greater plant K content might affect some physiological processes under stress. Specifically, the questions addressed were whether a greater K vacuolar accumulation might contribute to plant water acquisition and use for increasing productivity, and also if enhanced K capture might reduce water loss and provide greater protection of photosynthetic activity, resulting in increased water-use efficiency.

2. Materials and Methods

2.1. Plant Material and Stress Treatments

Two tomato lines, the cultivar MicroTom (henceforth, the wild-type, WT) [14] and a transgenic line (N367), were used. The line N367 overexpresses the vacuolar AtNHX1 cation/proton antiporter and is able to accumulate greater K quantities in vacuoles and tissues [13]. Seeds were pregerminated (28 °C, 4 days) and the seedlings were transplanted into quartz sand (0.5 L pots) and irrigated every three days with a modified Hewitt's nutrient solution supplemented with K at the concentrations required for each experiment. The basic nutrient solution contained (in mM): NO_3^-, 4; $H_2PO_4^-$, 1; SO_4^{2-}, 3; Ca^{2+}, 2; Mg^{2+}, 1; and (in μM): Fe, 50 (as FeEDDHA); Mn, 10; Cu, 1; Zn, 5; B, 30. Three different K concentrations were used as nutritional treatments by supplementing the basic solution with 0.1, 1 and 10 mM K (as K_2SO_4). When 0.1 mM K was added, 1 mM phosphate was provided as NaH_2PO_4. For addition of 10 mM K, Mg concentration in the solution was increased to 5 mM to avoid Mg deficiency [9]. Plants were grown in a greenhouse (mean temperature, 26 °C; mean relative humidity, 45.0%).

When plants reached the first flowering stage, water stress treatments were initiated by withholding irrigation during 2 weeks followed by one day with abundant watering (to flush pot sand) with nutrient solution supplemented with K as required. This treatment was repeated several times, depending on the experiment (Supplementary Materials Figure S1). When measuring yield, fruits were collected after six cycles of drying/watering, while control plants kept a regular watering regime every 3 days. To test the effect of acclimation, plants were treated with either one or four cycles of watering/drying. Plants submitted to a single water stress cycle were considered "nonacclimated", whereas plants that experienced four consecutive watering/drying cycles were judged as "acclimated". This preconditioning (priming) to stress allows osmotic adjustment and expression of drought-tolerance mechanisms [15,16]. Physiological measurements (leaf water status, photosynthesis, stomatal conductance) were taken at the end of the drying periods and before rewatering, as stated in the Results section. Control plants were irrigated normally every three days.

2.2. Leaf Water Status

Leaf water status was estimated by measuring relative water content (RWC) and leaf water content (LWCT) [17]. Leaf osmotic potentials were measured in leaf samples by using a thermocouple psychrometer (HR33 microvoltmeter and C-52 sample chamber, Wescor, Logan, UT, USA).

2.3. Gas Exchange and Photochemical Reflectance Index

Gas exchange measurements were performed in the youngest fully expanded leaves with an LC-Pro Photosynthesis System gas analyzer (ADC Bioscientific, Hoddesdon, UK). Leaf water-use efficiency (WUEi) was estimated as the ratio between photosynthesis (A) and stomatal conductance (gs). The photochemical reflectance index (PRI) was determined with a PlantPen PRI 200 (PSI, Drásov, Czech Republic).

2.4. Hydroponic Culture and Root Hydraulic Conductivity

For determining the effect of K supply on root hydraulic conductivity, a critical component of plant water uptake and transport, 1-week old seedlings of both MicroTom lines, WT and N367, were transferred into aerated nutrient solutions contained in 10-L plastic boxes as previously described [13] and cultivated hydroponically until flowering in the greenhouse.

Root hydraulic conductivity (L_0) was estimated by measuring root exudation rates in detopped plants [18] grown hydroponically with nutrient solutions containing different K concentrations [13]. Xylem saps exuded from roots during 1 h were collected in Eppendorf tubes with micropipettes and frozen at $-20\ °C$ for further analysis of K concentration.

2.5. Analysis of K Concentration

Potassium was determined in leaves used for porometry and gas exchange, as well as in remaining leaves, stems, roots and fruits, or in xylem saps from root hydraulic conductivity measurements. Plant material (leaves, stems, roots and fruits) was weighed before and after drying (48 h, 70 °C), and ground to powder. Potassium was extracted with distilled water by heating (90 °C, 1 h) and sonication, and measured by flame photometry (PFP7, Jenway, Staffordshire, UK). Potassium in xylem saps was measured with the same equipment after appropriate dilutions.

2.6. Harvest

Fruits were harvested after plants had received six watering/drying cycles starting immediately after the onset of flowering. At harvest, shoot, fruit and root fresh and dry weights and fruit number were recorded. Sugar contents in dried leaves, stems and roots were estimated after hot water extraction with the phenol–sulphuric method using a sucrose standard [19]. Soluble solids in fruits were estimated by refractometry as Brix degrees (Atago Hand held refractometer). Harvest index was calculated as the ratio among fruit weight and total plant weight. Potassium use efficiency (KUE) was calculated by dividing the fruit mass by the total plant K uptake.

2.7. Free Amino Acid Contents in Leaves and Flowers

Considering the effect of available K on plant amino acids concentration [20,21] and the protection provided by compatible solutes against drought stress [22], we measured free amino acids in leaves and flowers of water-stressed tomato under two K regimes. At full flowering stage, and after two cycles of watering/drying, fully expanded leaves and flowers from plants grown at 0.1 and 10 mM K were frozen in liquid N_2 and stored at $-70\ °C$. Then, they were homogenized in Eppendorf tubes with plastic pestles and centrifuged (10,000 rpm, 5 min at 4 °C) and free amino acids in the supernatants were separated and quantified after derivatization with phenylisotiocyanate by reversed-phase high-performance liquid chromatography [23].

2.8. Statistical Analysis

The experimental design was a complete randomized design with a factorial 3×2 (3 K concentrations, 2 water treatments) with 3–4 plants per treatment or a factorial 3×2 (3 K concentrations, 2 tomato lines) with four plants per treatment. Analysis of variance (ANOVA) was applied in order to examine significant differences between treatments (K levels, watering) among variables (plant growth parameters, gas exchange measurements, etc.). Normality of the data was checked by Shapiro–Wilks test. When differences were statistically significant, a post hoc test (Least Significant Difference, LSD) was applied for comparisons. A correlation analysis (Pearson) was done between the K concentration in leaves and gas exchange or PRI values, and between plant sugar and water content. Statistical analyses were performed with a standard package (Statistica, StatSoft Inc., Tulsa, OK, USA).

3. Results

To test the combined effect of K supply and water stress on the vegetative growth and yield, wild-type MicroTom plants were grown in sand and irrigated every 3 days with nutrient solution containing 0.1, 1 or 10 mM K. When plants reached the flowering stage, plants of each group were either irrigated normally or submitted to six consecutive cycles of drying and rewatering with nutrient solution (Supplementary Figure S1). Then, plants and fruits were harvested, and growth and yield parameters were measured. Results showed that water stress and the concentration of supplied K greatly affected fruit setting and therefore fruit yield (Table 1).

Table 1. Effect of K supply and watering treatment on plant growth and fruit production in tomato. Control plants (irrigated) were compared to plants submitted to six cycles of drying/watering after flower initiation (drought). Means followed by the same letter are not significantly different (LSD Test, $p < 0.05$). K and W in ANOVA represent potassium and water stress treatment factors, respectively.

K Supply (mM)	Water Treatment	Shoot (g)	Root (g)	Shoot/ Root Ratio	Number of Fruits	Fruit Size (g/fruit)	Yield (g/plant)
0.1	drought	4.4 ± 0.9 a	1.6 ± 0.4 a	2.7 ± 0.5 a	0	0	0
	irrigated	23.6 ± 3.4 b	3.4 ± 0.8 b	7.2 ± 1.4 b	7 ± 4.5 b	1.9 ± 1.2 ab	9.0 ± 3.5 b
1	drought	7.9 ± 1.1 c	1.9 ± 0.7 a	4.5 ± 1.4 c	2 ± 1.2 a	1.3 ± 0.8 b	2.0 ± 1.5 ac
	irrigated	34.2 ± 5.1 d	4.5 ± 1.3 c	7.9 ± 1.5 b	11 ± 1.6 c	1.9 ± 0.5 ab	20.4 ± 3.2 d
10	drought	9.6 ± 1.8 c	2.2 ± 0.9 a	4.8 ± 1.7 c	2 ± 1.7 a	1.7 ± 0.4 ab	3.3 ± 2.6 c
	irrigated	29.0 ± 3.2 e	4.1 ± 0.6 c	7.1 ± 0.9 b	9 ± 2.8 d	2.1 ± 0.5 a	17.2 ± 2.7 e
ANOVA							
K		$p < 0.001$	$p < 0.01$	$p < 0.01$	$p < 0.001$	$p < 0.001$	$p < 0.001$
W		$p < 0.001$	$p < 0.001$	$p < 0.001$	$p < 0.001$	$p < 0.001$	$p < 0.001$
K*W		$p < 0.001$	ns	$p < 0.05$	ns	$p < 0.01$	$p < 0.01$

Water stress after flowering significantly affected fruit yield but a greater K supply could partially counteract its negative impact on yield (Table 1). Plants growing in 0.1 K and submitted to water stress failed to produce fruits, and this could be prevented partially by increasing the K availability. However, when irrigation was restricted, the inhibition on fruit set had a much greater effect than K limitation. The same was true even when including line N367 (Table 2), which paradoxically showed a greater drought-induced reduction in fruit set and yield in comparison with the control line (WT) in spite of the higher K contents in the former (Table 3). Under regular irrigation, low K significantly reduced fruit set, and therefore, fruit yield. A surplus of K improved plant performance under drought by increasing shoot mass (greater photosynthetic area) while root mass was unaffected (Table 1).

Table 2. Effects of K supply on fruit yield, solute contents (Brix degrees) and fruit K concentration in two lines (wild-type (WT) and N367) differing in K uptake after six cycles of limited watering (see Section 2). Means followed by the same letter are not significantly different (LSD Test, $p < 0.05$). K and L in ANOVA represent potassium and line factors, respectively.

K Supply (mM)	Line	Fruit Number	Fruit Size (g/fruit)	Fruit Yield (g/plant)	Brix (°)	K (%)
0.1	WT	3 ± 1.1 ab	0.8 ± 0.7 a	1.8 ± 1.3 a	5.1 ± 0.4 a	21.6 ± 3.5 a
	N367	1 ± 0.6 a	0.2 ± 0.1 a	0.2 ± 0.1 a	5.9 ± 0.1 ab	27.9 ± 2.2 a
1	WT	5 ± 2.6 ab	2.5 ± 0.9 c	10.5 ± 2.2 b	5.9 ± 0.8 ab	32.3 ± 5.3 ab
	N367	6 ± 2.1 bc	1.0 ± 0.5 ab	6.0 ± 1.3 c	6.7 ± 0.8 b	43.0 ± 14.7 b
10	WT	7 ± 4.8 bc	1.9 ± 0.6 bc	11.0 ± 2.8 b	6.6 ± 0.9 b	60.5 ± 4.3 c
	N367	10 ± 2.8 c	0.9 ± 0.4 a	9.0 ± 3.1 bc	6.2 ± 0.9 b	60.6 ± 9.2 c
ANOVA						
K		$p < 0.01$	$p < 0.01$	$p < 0.001$	$p < 0.05$	$p < 0.001$
L		ns	$p < 0.01$	$p < 0.01$	ns	ns
K*L		ns	ns	ns	ns	ns

Table 3. Effects of K supply on plant growth and K concentration in different organs of two tomato lines (WT and N367) at harvest after receiving six cycles of limited watering. Means followed by the same letter are not significantly different (LSD Test, $p < 0.05$). K and L in ANOVA represent potassium and line factors, respectively. FW, fresh weight.

K Supply (mM)	Line	Growth (g FW/plant)			K Concentration (%)		
		Leaves	Stems	Roots	Leaves	Stems	Roots
0.1	WT	2.68 ± 0.32 a	2.20 ± 0.45 a	2.10 ± 0.35 ab	0.28 ± 0.10 a	0.38 ± 0.11 a	0.13 ± 0.07 a
	N367	2.30 ± 0.21 a	3.50 ± 0.28 b	1.48 ± 0.19 a	0.68 ± 0.15 ab	1.10 ± 0.10 a	0.28 ± 0.08 a
1	WT	3.48 ± 0.95 ab	2.85 ± 0.77 ab	2.86 ± 1.34 b	1.97 ± 0.51 b	2.70 ± 0.43 b	0.25 ± 0.02 a
	N367	2.85 ± 1.20 a	3.40 ± 0.64 b	1.80 ± 0.60 a	3.56 ± 0.46 c	4.29 ± 0.85 c	0.93 ± 0.27 a
10	WT	4.68 ± 1.64 b	3.68 ± 1.28 b	2.23 ± 0.67 ab	5.20 ± 1.57 d	5.95 ± 0.49 d	3.43 ± 0.49 b
	N367	3.70 ± 0.57 ab	3.47 ± 0.09 b	1.48 ± 0.14 a	4.44 ± 1.65 cd	8.00 ± 1.52 e	5.41 ± 1.80 c
ANOVA							
K		$p < 0.001$	ns	ns	$p < 0.001$	$p < 0.001$	$p < 0.001$
L		ns	ns	$p < 0.01$	ns	$p < 0.01$	$p < 0.01$
K*L		ns	ns	ns	ns	ns	Ns

The NHX1-overexpressing line N367 failed to improve the growth of leaves, stems or roots at harvest when compared with the control line (WT) in spite of accumulating more K than the WT (Table 3). A significant effect of K on fruit yield was found when K was increased from 0.1 to 1 mM K whereas a further increase in K supply had a marginal effect (Table 2). Surprisingly, N367 with the highest plant K contents (Table 3) showed a tendency towards producing more fruits than the WT as the K availability increased, but had smaller fruits and lower yield than the WT (Table 2). The ANOVA showed a "line" factor (WT vs N367) only significant for fruit size and yield, whereas the "K" factor was critical in all fruit characters measured (Table 2).

Increasing K supply significantly favored fruit K and soluble solid (Brix) accumulation in both lines at harvest (Table 2). While no differences in harvest index were found between lines, a significant lower K use efficiency was recorded for line N367 (Table 4).

Table 4. Differences in harvest index and K use efficiency between two tomato lines (WT and N367) grown with different K availabilities and subjected to six cycles of drying/watering. In columns, means followed by the same letter are not significantly different (LSD Test, $p < 0.05$). K and L in ANOVA represent potassium and line factors, respectively.

K Supply (mM)	Line	Harvest Index	K Use Efficiency
0.1	WT	1.4 a	13.1 a
	N367	1.0 a	3.4 b
1	WT	1.3 a	2.2 bc
	N367	0.8 a	0.7 c
10	WT	1.2 a	0.8 bc
	N367	1.1 a	0.5 c
	ANOVA		
	K	ns	$p < 0.001$
	L	$p < 0.1$	$p < 0.001$
	K*L	ns	$p < 0.001$

Next, we determined key physiological parameters in wild-type and N367 grown with 1 mM K in the nutrient solution until flowering (to avoid severe effects of the lowest K supply on growth) and then submitted to different regimes of K availability (0.1, 1 and 10 mM K). At flowering, both lines were submitted as in the previous experiment to six rounds of 2-week drought and one day of rewatering with the corresponding nutrient solution. In drought-treated wild-type plants, shoot and root water contents were improved by a greater supply of K (Table 5). The same trend was found for lines WT and N367, which at harvest showed greater water content in leaves (0.1 vs. 1–10 mM K) or in roots (0.1–1 vs. 10 mM K) (Table 6). However, stem water contents remained similar in spite of significant increases in their K concentration by increasing K supply (see Table 6).

Table 5. Effect of K supply and watering on shoot and root water content (g H_2O g dry weight^{-1}) in tomato at harvest. Control plants (irrigated) were compared to plants submitted to six cycles of drying/watering after flower initiation (drought). Means followed by the same letter are not significantly different (LSD Test, $p < 0.05$). K and W in ANOVA represent potassium and water factors, respectively.

K Supply (mM)	Water Treatment	Shoot Water Content	Root Water Content
0.1	drought	5.1 ± 0.7 a	3.0 ± 1.0 a
	irrigated	7.3 ± 0.6 b	6.3 ± 0.9 bc
1	drought	6.1 ± 0.5 c	4.8 ± 1.0 d
	irrigated	7.7 ± 0.4 b	6.9 ± 1.0 be
10	drought	6.6 ± 0.6 c	5.8 ± 0.8 b
	irrigated	7.5 ± 0.6 b	7.2 ± 0.8 e
ANOVA			
K		$p < 0.001$	$p < 0.01$
W		$p < 0.001$	$p < 0.001$
K*W		$p < 0.001$	$p < 0.01$

Plants receiving less K showed clear wilting symptoms at the end of each drought cycle in comparison with plants receiving a higher K supply (Figure 1).

At harvest, sugar accumulation in leaves was recorded in drought-treated WT and N367 tomato plants at the lowest K supply (Table 6) but no association was found between sugar accumulation and water content, with the exception of roots, which showed a significant correlation between their sugar and water contents ($r = 0.61, p < 0.001, n = 24$). The K concentration provided also had a significant effect on the water content of leaves, stems and roots (Table 6).

Table 6. Water and sugar contents in leaves, stems and roots of tomato WT and N367 lines at harvest when grown with different K concentrations and subjected to six cyclical drought periods. Means followed by the same letter are not significantly different (LSD Test, $p < 0.05$). K and L in ANOVA represent potassium and line factors, respectively.

K Supply (mM)		Water Content (g g dry wt^{-1})			Sugars (%)		
	Line	Leaves	Stems	Roots	Leaves	Stems	Roots
0.1	WT	4.5 ± 2.0 a	7.4 ± 1.3 a	5.9 ± 1.5 a	16.4 ± 4.7 ab	6.8 ± 1.1 a	3.6 ± 1.3 a
	N367	4.4 ± 1.3 a	6.4 ± 1.5 a	5.6 ± 1.4 a	20.4 ± 4.9 a	11.8 ± 2.3 bc	4.9 ± 1.5 a
1	WT	6.3 ± 0.3 b	5.9 ± 0.3 a	4.7 ± 1.5 a	12.5 ± 1.4 bc	13.1 ± 3.5 bd	4.7 ± 3.0 a
	N367	6.1 ± 1.5 b	6.1 ± 0.3 a	5.9 ± 0.7 a	13.1 ± 1.2 bc	10.6 ± 3.9 abc	4.6 ± 1.2 a
10	WT	7.2 ± 0.5 b	6.1 ± 0.6 a	9.2 ± 1.5 b	13.9 ± 3.5 bc	16.0 ± 3.2 d	11.9 ± 3.3 b
	N367	6.9 ± 0.5 b	7.1 ± 0.4 a	10.9 ± 2.0 b	8.9 ± 2.5 c	8.3 ± 1.2 ac	6.9 ± 3.7 a
	ANOVA						
	K	$p < 0.05$	ns	$p < 0.001$	$p < 0.001$	ns	$p < 0.001$
	Lines	ns	ns	$p < 0.05$	ns	ns	Ns
	K*L	$p < 0.05$	ns	ns	$p < 0.05$	$p < 0.001$	Ns

Figure 1. Plants of tomato lines WT and N367 growing with 0.1 (**left**), 1 (**center**) or 10 mM K (**right**) after withholding irrigation for two weeks.

The potassium supply also affected the ability for water acquisition by modifying hydraulic conductivity (L_0) in hydroponically grown plants. Root L_0 values showed significant differences among K treatments (Table 7). In low 0.1 mM K, root L_0 was greater than at higher K supplies. Greater L_0 values correlated with greater sap flows (J_v) but lower K contents in the xylem sap, indicating that plants absorbed more water at low-K supply to satisfy nutritional requirements. Significant differences between lines were also recorded at 0.1 mM K (Table 7). N367 plants had a higher K content in their xylem sap compared to controls when grown in sufficient K (1 and 10 mM). They also showed reduced sap flows (J_v) and hydraulic conductivity (L_0) at low 0.1 mM K.

Both water stress and K availability alter the concentration of amino acids in plants [19,20], some of which, like proline, have protective effects under drought stress [21]. Hence, we measured free amino acids in the leaves and flowers of water-stressed tomato plants submitted to two cycles of watering/drying under 0.1 and 10 mM K regimes. Indeed, nitrogen metabolism in MicroTom was also affected by K nutrition and water supply, as found in the variation in the free amino acids pool in leaves and flowers. At low K, the leaves from plants suffering water stress accumulated mostly arginine and proline (Supplementary Figure S2a) whereas at higher K supply, leaf proline content was significantly increased while arginine content remained rather unaffected. In flowers, glutamine, asparagine, proline and arginine were the main amino acids stored in both irrigated and water-stressed plants and supplying more K increased the concentration of them all (Supplementary Figure S2b).

Table 7. Effect of available K on sap flow (J_v), root hydraulic conductivity (L_0) and xylem sap K concentration in hydroponically grown tomato lines WT and N367 at flowering stage. Means followed by the same letter are not significantly different (LSD Test, $p < 0.05$). K and L in ANOVA represent potassium and line factors, respectively.

K (mM)	Line	J_v	L_0	Sap K
0.1	WT	216.3 ± 24.6 a	1221.9 ± 138.8 a	3.5 ± 0.97 a
	N367	170.3 ± 19.2 b	962.2 ± 108.2 b	5.0 ± 1.1 a
1	WT	60.3 ± 13.4 c	370.2 ± 82.0 c	26.7 ± 2.3 b
	N367	56.6 ± 18.0 c	347.2 ± 110.6 c	29.4 ± 6.4 bc
10	WT	78.2 ± 19.6 c	355.2 ± 88.9 c	30.9 ± 2.1 c
	N367	82.2 ± 11.6 c	373.6 ± 52.8 c	35.7 ± 0.9 d
	ANOVA			
	K	$p < 0.001$	$p < 0.001$	$p < 0.001$
	Lines	$p < 0.05$	$p < 0.05$	$p < 0.05$
	K*Lines	$p < 0.05$	$p < 0.05$	Ns

Units: J_v, sap flow, mg·g root fresh wt^{-1} h^{-1}; L_0, root hydraulic conductivity, mg·g root fresh wt^{-1} h^{-1} MPa^{-1}; sap K, K concentration in the xylem sap, mM.

Priming plants with stress episodes improves plant adaptability to subsequent and repetitive stresses in a process known as acclimation, which enhances fitness and yield [15,16]. Hence, we tested how repetitive episodes of drying/watering (priming) affected the photosynthesis parameters, gas exchange and water-use efficiency of MicroTom plants at varying K availabilities. After reaching the flowering stage, plants growing with 0.1, 1 and 10 mM in the nutrient solution were submitted to either one cycle or four consecutive 2-week cycles of water withholding and rewatering (see Section 2). Plants experiencing only one water stress cycle were considered "nonacclimated", whereas primed plants that experienced three additional watering/drying cycles were deemed to be "acclimated". Physiological parameters were measured at the end of the one or four cycles of drying/watering and compared to the control plants that had been irrigated normally. The irrigation of control and treated plants was with complete nutrient solutions containing different K levels (0.1, 1 and 10 mM K). As expected, low water availability significantly reduced photosynthesis in acclimated and nonacclimated plants alike (Table 8). In plants suffering water stress, leaf photosynthetic and transpiration rates were limited by stomatal closure, and the increase in K supply did not prevent the drought-induced photosynthetic inhibition. However, increasing K availability to plants led to an improvement in intrinsic water-use efficiency (WUEi) in both irrigated and water-stressed plants (Table 8). Leaf K concentration was negatively associated with transpiration, stomatal conductance and photosynthesis ($r = -0.54$, $r = -0.51$, $r = -0.46$, $p < 0.001$, $n = 36$) but it was positively correlated with water-use efficiency ($r = 0.50$, $p < 0.001$, $n = 36$). The only positive association of leaf K contents was found in the photochemical reflectance index (PRI) ($r = 0.61$, $p < 0.001$, $n = 44$) at low K supply (see Supplementary Figure S3). At higher K supplies, leaf K contents were weakly and negatively associated with PRI ($r = -0.27$, $p < 0.05$, $n = 80$). When plants were grown with unlimited water supply, no association was found between K supply and stomatal conductance or transpiration rate ($r = 0.06$ and $r = 0.09$, $n = 30$). Priming plants led to an apparent improvement in the efficiency of C gain per mole of stomatal water lost, i.e., a greater increase in WUEi, in comparison with the irrigated controls (Table 8).

Table 8. Photosynthesis (*A*), transpiration rate (*E*), stomatal conductance (g_s) and intrinsic water-use efficiency (WUE$_i$) in plants that were watered every three days (irrigated), or received one cycle of drying/watering (nonacclimated) or four cycles (acclimated). Photosynthesis and gas exchange were measured after one or the four cycles of drying/watering, in comparison with plants that were irrigated normally. In the irrigation, plants had different K regimes (0.1, 1 and 10 mM K). Within each column, means (±standard deviation) followed by the same letter are not significantly different (LSD Test, $p < 0.05$). Units: *A*, µmol CO_2 m^{-2} s^{-1}; *E*, mmol H_2O m^{-2} s^{-1}; g_s, mmol m^{-2} s^{-1}; WUE$_i$, µmol $CO_2 \cdot$mmol^{-1}. K and W in ANOVA represent potassium and watering regime, respectively.

K Supply (mM)	Water Treatment	*A*	*E*	g_s	WUE$_i$
		Nonacclimated plants			
0.1	drought	10.3 ± 0.8 a	5.4 ± 0.1 a	0.18 ± 0.01 a	56.4 ± 7.5 a
	irrigated	16.7 ± 0.8 b	11.2 ± 0.7 b	0.89 ± 0.09 b	20.8 ± 6.0 b
1	drought	7.5 ± 2.6 c	2.4 ± 0.7 c	0.09 ± 0.03 a	87.4 ± 7.5 c
	irrigated	15.2 ± 0.6 b	8.0 ± 0.5 d	0.40 ± 0.06 c	40.5 ± 4.3 d
10	drought	9.9 ± 2.5 a	4.5 ± 1.1 a	0.15 ± 0.05 a	67.1 ± 8.5 a
	irrigated	15.9 ± 1.9 b	7.5 ± 1.4 d	0.35 ± 0.10 c	47.9 ± 10.7 ad
ANOVA					
K		Ns	$p < 0.001$	$p < 0.001$	$p < 0.01$
W		$p < 0.001$	$p < 0.001$	$p < 0.001$	$p < 0.01$
K*W		Ns	$p < 0.01$	$p < 0.001$	$p < 0.01$
		Acclimated plants			
0.1	drought	5.1 ± 2.6 ac	1.3 ± 0.6 a	0.07 ± 0.04 a	81.8 ± 40.6 a
	irrigated	8.3 ± 1.3 b	2.2 ± 0.5 b	0.12 ± 0.05 b	75.8 ± 21.3 a
1	drought	3.4 ± 1.2 c	0.6 ± 0.3 c	0.02 ± 0.02 c	214.9 ± 126.4 b
	irrigated	5.5 ± 2.1 ac	1.1 ± 0.5 a	0.05 ± 0.02 ac	128.9 ± 60.6 ad
10	drought	5.8 ± 3.5 abc	0.8 ± 0.3 ac	0.03 ± 0.01 ac	164.4 ± 52.3 b
	irrigated	6.4 ± 1.9 ab	1.0 ± 0.2 ac	0.04 ± 0.01 ac	148.3 ± 31.7 abd
ANOVA					
K		$p < 0.05$	$p < 0.001$	$p < 0.001$	$p < 0.01$
W		$p < 0.001$	$p < 0.001$	$p < 0.001$	ns
K*W		Ns	ns	ns	ns

4. Discussion

In this work we have reassessed the contribution of K supply towards key physiological parameters in tomato under water stress with the distinctive advantage of comparing two isogenic lines differing in their ability to take up and compartmentalize K into vacuoles [13]. We found that high K in the nutrient solution improved the water contents in wild-type plants, particularly under limited watering conditions (Tables 5 and 6). A significant beneficial effect of K supplementation was also found in fruit number and yield in MicroTom plants (Tables 1 and 2). The yield of well-irrigated wild-type plants doubled with K supplement in the 1–10 mM range compared to 0.1 mM, and the increase was five-fold in water-stressed plants. However, the yield was most severely impaired by drought and this could not be overcome by K supplements since the yield of stressed plants remained five- to ten-fold lower than that of control plants at 1 and 10 mM K (Table 1). The benefit of K supplements was more intense in the transgenic line N367 under limited watering, although the yield of N367 plants approached that of the wild-type only at 10 mM, the highest concentration tested (Table 2). This differential behavior of N367 plants is likely related to the larger proportion of cellular K that is compartmentalized inside vacuoles in the transgenic line at the expense of the cytosolic pool [13], so that a higher K supplement is needed to off-set the enhanced compartmentation. In other words, greater K compartmentation into vacuoles may be detrimental when low K availability and water stress are combined.

The enhanced accumulation of K in vacuoles, achieved by the overexpression of tonoplast- and endosomal-localized NHX antiporters, has been shown to the increase the salt tolerance of tomato plants by improving K homeostasis and reducing salt-induced K loss [13,24,25]. However, we show that the tomato line N367 with greater K accumulation failed to demonstrate a better performance under water stress compared to wild-type when plants suffered from water restrictions in spite of the consistently higher K contents in N367 plants (Tables 2 and 3). In the line N367, K sequestration in stems and roots (see Table 3) probably limited the nutrients available for assimilate redistribution to developing sinks (i.e., fruits) thereby reducing its K use efficiency (Table 4). However, the line N367 accumulated a significant amount of sugars in the leaves and stems at low K (Table 6). Sugars may accumulate either in response to water stress [26] or as a symptom of K deficiency [27,28]. The line N367 accumulates sugars under salt-stress or limited K supply [13,21], and yet it shows greater sensitivity to K deficiency in spite of similar water contents compared to wild-type MicroTom plants (Table 6). Sugar accumulation in N367, which results in lower assimilate transport to sinks, might explain the reduced fruit set and size and fruit yield (Table 2).

The limitation in leaf transpiration imposed by water scarcity led to a significant improvement in WUE at optimum K nutrition, mostly in acclimated plants, by reducing more water loss than photosynthetic carbon gain. Increasing K also significantly affected water economy by favoring stomatal closure (reducing water loss) under water stress and by reducing root hydraulic conductivity (with no water restrictions, see Table 7). However, greater K supply did not protect the photosynthetic rate under water stress as previously shown in wheat [3,4]. Only when the external K concentration was low, the leaf K contents were associated with an increase in PRI, i.e., leaf K might have a protective effect on the photosynthetic apparatus under drought [29].

High K improved carbon gain under water stress by increasing photosynthetic area, i.e., promoting shoot growth and therefore leaf expansion (Tables 1 and 3) and probably through the increased synthesis of protective osmolytes like proline (Supplementary Figure S2a). In our controlled conditions, K improved the shoot/root ratio (Table 1) and leaf WUE (Table 8) in water-stressed plants but it led to marginal increase in fruit yield (Table 2). The improvement in shoot growth and water-use efficiency by increasing K has also been found in other species (albeit by using different approaches [30–32]). The role of K in the osmotic potential and turgor generation required for expansive growth [1,33] may explain the significant increase in shoot growth of either water-deprived plants or irrigated plants when K supply was increased (Table 3).

The greater root hydraulic conductivity found at low K in nutrient solutions (Table 7) might be explained by K-deficiency sensing, which causes root ABA accumulation, increases in radial water flow and facilitates root water uptake [34–37]. Potassium deficiency also causes ethylene release and therefore ethylene-induced defective stomatal closure cannot be discarded [38,39] in the observed higher stomatal conductance and transpiration rate at low K (Table 8). As K deficiency progresses, root water conductivity diminishes by inhibition of aquaporins and K-uptake channels [12] and the accumulation of ABA in leaves may limit the effects of ethylene on stomata [40] leading to stomatal closure. On the other hand, increasing K supply may reduce root hydraulic conductance (Table 7), which might be responsible, at least partially, for the decrease in stomatal conductance (Table 8), as was found for other species [18].

The increased leaf proline contents under high K (Supplementary Figure S2a) may have provided both osmotic adjustment and radical scavenging to protect metabolic processes under water stress [22,41], leading to increased fruit setting and growth (Table 3). In flowers, the small changes in amino acid concentrations recorded either under water or K shortages (Supplementary Figure S2b) show them as preferential sinks and therefore less affected by stress [42].

A strong association between plant water content and K uptake and use is evident from the experiments shown here and in the related literature [11,12]. Undoubtedly, K contributes to maintaining plant water status through its effects on the regulation of root water conductivity and stomatal movements. It improves photosynthetic carbon assimilation under water scarcity, and thereby fruit yield, by improving shoot growth, but had no effect on photosynthetic rates. However, the yield penalty caused by water limitation cannot be overcome by increasing the K supply. By using a tomato line (N367) able to accumulate higher K concentrations, the yield was not improved but in fact reduced by limiting K use efficiency (Table 4). The potassium stored in vegetative parts improved plant water contents but was a deterrent for getting better yields under drought as it was not available for the transport of assimilates into fruits.

When water and nutrient limitation affects different crops, the relationship between K availability and yield is not a simple one and the yield responses depend on several factors inherent to each species [43] and crop nutrient use efficiency [44]. In tomato, K may affect water acquisition and economy as well as C gain, and its partitioning in a close interplay and nutrient use efficiency is of paramount importance for fruit yield. A greater K accumulation (N367) was not translated into an improvement of plant growth or fruit yield when plants suffered water stress, but indeed proved to be a disadvantage by reducing plant growth and consequently fruit setting and size.

In conclusion, K supplementation had a protective effect on the yield of tomato plants under limited watering, but this benefit was insufficient to off-set the yield loss induced by water shortage. Increasing the plant K contents through enhanced compartmentation into vacuoles failed to protect yield under drought stress, even though it could improve some physiological traits.

Supplementary Materials: The following is available online at https://www.mdpi.com/2311-7524/7/2/20/s1. Figure S1: Scheme showing the different regimes of drying/rewatering used to measure the effect of K availability on the yield and acclimation of water-stressed plants. Figure S2: Free amino acids in (a) leaves and (b) flowers from tomato plants with or without water stress when grown at low (0.1 mM K) or high K (10 mM K). Figure S3: Effect of K and watering treatments on (a) leaf K concentration and (b) photochemical reflectance index in two tomato lines (WT and N367).

Author Contributions: E.O.L. designed and supervised the research; A.D.L., M.C. (Mireia Corell), M.C. (Mathilde Chivet) performed experiments and analyses; all authors analyzed and discussed the data; E.O.L. and J.M.P. wrote the manuscript and obtained the funding. All authors have read and agreed to the published version of the manuscript.

Funding: This work was supported by grant RTI2018-094027-B-I00 from the Spanish Agencia Estatal de Investigación (AEI), of Ministerio de Ciencia, Innovación y Universidades (MCIU), and cofinanced by the European Regional Development Fund, to E.O.L. and J.M.P.

Institutional Review Board Statement: Not applicable.

Informed Consent Statement: Not applicable.

Data Availability Statement: Data supporting the present results are available on request to the corresponding author.

Acknowledgments: The authors thank R. Fernández, A. Cifarelli and D.M. Cabrera for their contributions to different experiments while in periods of training, and Francisco J. Quintero and Anna M. Lindahl for critical reading of the manuscript. F.J. Quintero produced the scheme in Supplementary Figure S1.

Conflicts of Interest: The authors declare no conflict of interest.

References

1. Hawkesford, M.; Horst, W.; Kichey, T.; Lambers, H.; Schjoerring, J.; Skrumsager Møller, I.; White, P. Functions of macronutrients. In *Marschner's Mineral Nutrition of Higher Plants*, 3rd ed.; Marschner, P., Ed.; Elsevier Ltd.: Amsterdam, The Netherlands, 2012; pp. 135–189. [CrossRef]
2. Cakmak, I. The role of potassium in alleviating detrimental effects of abiotic stresses in plants. *J. Plant Nutr. Soil Sci.* **2005**, *168*, 521–530. [CrossRef]
3. Pier, P.A.; Berkowitz, G.A. Modulation of water stress effects on photosynthesis by altered leaf K+. *Plant Physiol.* **1987**, *85*, 655–661. [CrossRef]
4. Sen Gupta, A.; Berkowitz, G.A.; Pier, P.A. Maintenance of photosynthesis at low leaf water potential in wheat. Role of potassium status and irrigation history. *Plant Physiol.* **1989**, *89*, 1358–1365. [CrossRef] [PubMed]

5. Liebersbach, H.; Steingrobe, B.; Claassen, N. Roots regulate ion transport in the rhizosphere to counteract reduced mobility in dry soil. *Plant Soil* **2004**, *260*, 79–88. [CrossRef]

6. Tanner, W.; Beevers, H. Transpiration, a prerequisite for long-distance transport of minerals in plants? *Proc. Natl. Acad. Sci. USA* **2001**, *98*, 9443–9447. [CrossRef] [PubMed]

7. White, P.J.; Karley, A.J. Potassium. In *Cell Biology of Metals and Nutrients*; Hell, R., Mengel, R.R., Eds.; Springer: Berlin/Heidelberg, Germany, 2010; pp. 199–224. [CrossRef]

8. Ragel, P.; Raddatz, N.; Leidi, E.O.; Quintero, F.J.; Pardo, J.M. Regulation of K+ Nutrition in Plants. *Front. Plant Sci.* **2019**, *10*, 281. [CrossRef]

9. Sainju, U.M.; Dris, R.; Singh, B. Mineral nutrition of tomato. *Food Agric. Environ.* **2003**, *1*, 176–183.

10. Besford, R.T.; Maw, G.A. Effect of potassium nutrition on tomato plant growth and fruit development. *Plant Soil* **1975**, *42*, 395–412. [CrossRef]

11. Del Amor, F.M.; Marcelis, L.F. Regulation of K uptake, water uptake, and growth of tomato during K starvation and recovery. *Sci. Hortic.* **2004**, *100*, 83–101. [CrossRef]

12. Kanai, S.; Moghaieb, R.E.; El-Shemy, H.A.; Panigrahi, R.; Mohapatra, P.K.; Ito, J.; Nguyen, N.T.; Saneoka, H.; Fujita, K. Potassium deficiency affects water status and photosynthetic rate of the vegetative sink in green house tomato prior to its effects on source activity. *Plant Sci.* **2011**, *180*, 368–374. [CrossRef]

13. Leidi, E.O.; Barragán, V.; Rubio, L.; El-Hamdaoui, A.; Ruiz, M.T.; Cubero, B.; Fernández, J.A.; Bressan, R.A.; Hasegawa, P.M.; Quintero, F.J.; et al. The AtNHX1 exchanger mediates potassium compartmentation in vacuoles of transgenic tomato. *Plant J.* **2010**, *61*, 495–506. [CrossRef] [PubMed]

14. Meissner, R.; Jacobson, Y.; Melamed, S.; Levyatuv, S.; Shalev, G.; Ashri, A.; Elkind, Y.; Levy, A. A new model system for tomato genetics. *Plant J.* **1997**, *12*, 1465–1472. [CrossRef]

15. Yordanov, I.; Velikova, V.; Tsonev, T. Plant Responses to Drought, Acclimation, and Stress Tolerance. *Photosynthetica* **2000**, *38*, 171–186. [CrossRef]

16. E Menezes-Silva, P.; Sanglard, L.M.V.P.; Ávila, R.T.; E Morais, L.; Martins, S.C.V.; Nobres, P.; Patreze, C.M.; A Ferreira, M.; Araújo, W.L.; Fernie, A.R.; et al. Photosynthetic and metabolic acclimation to repeated drought events play key roles in drought tolerance in coffee. *J. Exp. Bot.* **2017**, *68*, 4309–4322. [CrossRef] [PubMed]

17. Reigosa Roger, M.J. *Handbook of Plant Ecophysiology Techniques*; Kluwer Academic Publishers: Dordrecht, The Netherland, 2001; ISBN 978-0-306-48057-7.

18. Cabañero, F.J.; Carvajal, M. Different cation stresses affect specifically osmotic root hydraulic conductance, involving aquaporins, ATPase and xylem loading of ions in Capsicum annuum, L. plants. *J. Plant Physiol.* **2007**, *164*, 1300–1310. [CrossRef] [PubMed]

19. Ashwell, G. New colorimetric methods of sugar analysis. In *Methods in Enzymology*; Neufeld, E.F., Ginsburg, V., Eds.; Academic Press: New York, NY, USA, 1969; Volume 8, pp. 85–95.

20. Armengaud, P.; Sulpice, R.; Miller, A.J.; Stitt, M.; Amtmann, A.; Gibon, Y. Multilevel Analysis of Primary Metabolism Provides New Insights into the Role of Potassium Nutrition for Glycolysis and Nitrogen Assimilation in Arabidopsis Roots. *Plant Physiol.* **2009**, *150*, 772–785. [CrossRef] [PubMed]

21. De Luca, A.; Pardo, J.M.; Leidi, E.O. Pleiotropic effects of enhancing vacuolar K/H exchange in tomato. *Physiol. Plant.* **2017**, *163*, 88–102. [CrossRef]

22. Verslues, P.E.; Sharma, S. Proline Metabolism and Its Implications for Plant-Environment Interaction. *Arab. Book* **2010**, *8*, e0140. [CrossRef]

23. Heinrikson, R.L.; Meredith, S.C. Amino acid analysis by reverse-phase high-performance liquid chromatography: Precolumn derivatization with phenylisothiocyanate. *Anal. Biochem.* **1984**, *136*, 65–74. [CrossRef]

24. Rodriguez-Rosales, M.P.; Jiang, X.; Gálvez, F.J.; Aranda, M.N.; Cubero, B.; Venema, K. Overexpression of the tomato K+/H+ antiporter LeNHX2 confers salt tolerance by improving potassium compartmentalization. *New Phytol.* **2008**, *179*, 366–377. [CrossRef]

25. Huertas, R.; Rubio, L.; Cagnac, O.; García-Sánchez, M.J.; Alché, J.D.D.; Venema, K.; Fernández, J.A.; Rodríguez-Rosales, M.P. The K+/H+antiporter LeNHX2 increases salt tolerance by improving K+homeostasis in transgenic tomato. *Plant Cell Environ.* **2013**, *36*, 2135–2149. [CrossRef] [PubMed]

26. Hanson, A.D.; Hitz, W.D. Metabolic Responses of Mesophytes to Plant Water Deficits. *Annu. Rev. Plant Physiol.* **1982**, *33*, 163–203. [CrossRef]

27. Mengel, K.; Viro, M. Effect of Potassium Supply on the Transport of Photosynthates to the Fruits of Tomatoes (Lycopersicon esculentum). *Physiol. Plant.* **1974**, *30*, 295–300. [CrossRef]

28. Sung, J.; Sonn, Y.; Lee, Y.; Kang, S.; Ha, S.; Krishnan, H.B.; Oh, T.-K. Compositional changes of selected amino acids, organic acids, and soluble sugars in the xylem sap of N, P, or K-deficient tomato plants. *J. Plant Nutr. Soil Sci.* **2015**, *178*, 792–797. [CrossRef]

29. Zhang, C.; Filella, I.; Liu, D.; Ogaya, R.; Llusia, J.; Asensio, D.; Peñuelas, J. Photochemical Reflectance Index (PRI) for Detecting Responses of Diurnal and Seasonal Photosynthetic Activity to Experimental Drought and Warming in a Mediterranean Shrubland. *Remote. Sens.* **2017**, *9*, 1189. [CrossRef]

30. Andersen, M.N.; Jensen, C.R.; Losch, R. The Interaction Effects of Potassium and Drought in Field-Grown Barley. I. Yield, Water-Use Efficiency and Growth. *Acta Agric. Scand. Sect. B Plant Soil Sci.* **1992**, *42*, 34–44. [CrossRef]

31. Egilla, J.N.; Davies, F.T.; Boutton, T.W. Drought stress influences leaf water content, photosynthesis, and water-use efficiency of Hibiscus rosa-sinensis at three potassium concentrations. *Photosynthetica* **2005**, *43*, 135–140. [CrossRef]
32. Jákli, B.; Tränkner, M.; Senbayram, M.; Dittert, K. Adequate supply of potassium improves plant water-use efficiency but not leaf water-use efficiency of spring wheat. *J. Plant Nutr. Soil Sci.* **2016**, *179*, 733–745. [CrossRef]
33. Wang, M.; Zheng, Q.; Shen, Q.; Guo, S. The Critical Role of Potassium in Plant Stress Response. *Int. J. Mol. Sci.* **2013**, *14*, 7370–7390. [CrossRef]
34. Quintero, J.M.; Fournier, J.M.; Ramos, J.; Benlloch, M. K+ status and ABA affect both exudation rate and hydraulic conductivity in sunflower roots. *Physiol. Plant.* **1998**, *102*, 279–284. [CrossRef]
35. Hose, E.; Steudle, E.; Hartung, W. Abscisic acid and hydraulic conductivity of maize roots: A study using cell- and root-pressure probes. *Planta* **2000**, *211*, 874–882. [CrossRef]
36. Schraut, D.; Heilmeier, H.; Hartung, W. Radial transport of water and abscisic acid (ABA) in roots of Zea mays under conditions of nutrient deficiency. *J. Exp. Bot.* **2005**, *56*, 879–886. [CrossRef] [PubMed]
37. Jiang, F.; Hartung, W. Long-distance signalling of abscisic acid (ABA): The factors regulating the intensity of the ABA signal. *J. Exp. Bot.* **2007**, *59*, 37–43. [CrossRef] [PubMed]
38. Tanaka, Y.; Sano, T.; Tamaoki, M.; Nakajima, N.; Kondo, N.; Hasezawa, S. Ethylene Inhibits Abscisic Acid-Induced Stomatal Closure in Arabidopsis. *Plant Physiol.* **2005**, *138*, 2337–2343. [CrossRef] [PubMed]
39. Benlloch-González, M.; Romera, J.; Cristescu, S.; Harren, F.J.M.; Fournier, J.M.; Benlloch, M. K+ starvation inhibits water-stress-induced stomatal closure via ethylene synthesis in sunflower plants. *J. Exp. Bot.* **2010**, *61*, 1139–1145. [CrossRef]
40. Sharp, R.E.; Lenoble, M.E.; Else, M.A.; Thorne, E.T.; Gherardi, F. Endogenous ABA maintains shoot growth in tomato independently of effects on plant water balance: Evidence for an interaction with ethylene. *J. Exp. Bot.* **2000**, *51*, 1575–1584. [CrossRef]
41. Sperdouli, I.; Moustakas, M. Interaction of proline, sugars, and anthocyanins during photosynthetic acclimation of Arabidopsis thaliana to drought stress. *J. Plant Physiol.* **2012**, *169*, 577–585. [CrossRef]
42. Borghi, M.; Fernie, A.R. Floral metabolism of sugars and aminoacids: Implications for pollinators' preferences and seed and fruit set. *Plant Physiol.* **2017**, *175*, 1510–1524. [CrossRef]
43. Schilling, G.; Eißner, H.; Schmidt, L.; Peiter, E. Yield formation of five crop species under water shortage and differential potassium supply. *J. Plant Nutr. Soil Sci.* **2016**, *179*, 234–243. [CrossRef]
44. Rengel, Z.; Damon, P.M. Crops and genotypes differ in efficiency of potassium uptake and use. *Physiol. Plant.* **2008**, *133*, 624–636. [CrossRef]

horticulturae

Article

Novel *S. pennellii* × *S. lycopersicum* Hybrid Rootstocks for Tomato Production with Reduced Water and Nutrient Supply

Jan Ellenberger [1], Aylin Bulut [1], Philip Blömeke [1] and Simone Röhlen-Schmittgen [1,2,*]

[1] Department of Horticultural Sciences, Institute of Crop Science and Resource Conservation (INRES), University of Bonn, 53119 Bonn, Germany; ellenberger@uni-bonn.de (J.E.); s7aybulu@uni-bonn.de (A.B.); philip.bloemeke@uni-bonn.de (P.B.)

[2] Department of Vegetable Crops, Hochschule Geisenheim University, 65366 Geisenheim, Germany

* Correspondence: Simone.RoehlenSchmittgen@hs-gm.de

Abstract: Drought stress and nutrient deficiency are limiting factors in vegetable production that will have a decisive role due to the challenges of climate change in the future. The negative effects of these stressors on yield can be mitigated by crop grafting. The increasing demands for resource-use efficiency in crop production, therefore, require the development and phenotyping of more resilient rootstocks, and the selection of appropriate scions. We tested the effect of combined drought stress and nutrient deficiency on yield and fruit quality of the two tomato cultivars 'Lyterno' and 'Tastery' in the greenhouse, grafted onto different rootstock genotypes. The use of four different rootstocks, including two novel *S. pennellii* × *S. lycopersicum* hybrids and the proven-effective use of 'Beaufort', as well as self-grafted plants, allowed conclusions to be drawn about the differential stress mitigation of the rootstocks used. The stress-induced yield reduction of the scion 'Lyterno' can be mitigated more significantly by the novel hybrid rootstocks than by the commercial rootstock 'Beaufort'. At the same time, however, the individual fruit weight and the lycopene content of the fruits were significantly reduced when grafted onto the hybrid rootstocks. In contrast, the cultivar 'Tastery' showed a weak stress response, so that a generally positive influence of the rootstocks independently of the scions could not be demonstrated. We conclude that, particularly for more sensitive cultivars, the selection of more resilient rootstocks offers the potential for sustainable and resource-efficient production not competing with the overall quality of tomatoes.

Keywords: grafting; water-use efficiency; nutrient use efficiency; vegetable production

Citation: Ellenberger, J.; Bulut, A.; Blömeke, P.; Röhlen-Schmittgen, S. Novel *S. pennellii* × *S. lycopersicum* Hybrid Rootstocks for Tomato Production with Reduced Water and Nutrient Supply. *Horticulturae* 2021, 7, 355. https://doi.org/10.3390/horticulturae7100355

Academic Editor: Antonio Ferrante

Received: 27 August 2021
Accepted: 17 September 2021
Published: 2 October 2021

Publisher's Note: MDPI stays neutral with regard to jurisdictional claims in published maps and institutional affiliations.

Copyright: © 2021 by the authors. Licensee MDPI, Basel, Switzerland. This article is an open access article distributed under the terms and conditions of the Creative Commons Attribution (CC BY) license (https://creativecommons.org/licenses/by/4.0/).

1. Introduction

Among abiotic stresses, drought is the most critical threat to agricultural production. It is one of the most common limiting factors influencing plant growth and development [1]. Agronomists are already using a range of management tools to cope with dry conditions in order to harvest sufficient yields at reasonable water and nutrient input [2]. At the same time, both commercial and public organizations across the world continue to develop new crop varieties to cope with the changing environment. Now, as climate change continues, both the physical environment in which agriculture is practiced and the market environment, as well as consumer demands are changing. Customers in parts of Europe are increasingly demanding not only regional products, but also those with a low energy and resource footprint [3]. Modern crop breeding has to keep up with these changing demands. As a result, breeding techniques for resilience and resource-use efficiency are investigated [4]. Different wild crops were included to stabilize yield. Nowadays many tomatoes carry the genetic information of related wild plants [5]. There are many positive known aspects from wild plants to be used. *S. pennellii*, as a tomato plant example, consists of genes of arid-adapted plant species from South America. *S. pennellii* can be crossed with tomato [6], but as its fruits are not edible, using *S. pennellii*—based accessions as a stress-tolerant rootstock seems to be more promising. Although it is possible to use gene

21

editing to create de novo plants with beneficial root systems and edible fruits at the same time [7], the lack of public acceptance will likely prevent the commercial use of such plants in Europe in the coming years and perhaps decades [8,9]. Nevertheless, the grafting of horticultural crops remains a method to improve plant growth and yield production [10,11]. Indeed, grafting can be helpful in combining two advantageous characteristics (rootstock and scion) in a single plant. For example, the grafting method is used and researched in vegetable crops to increase yield, fruit quality, and nutrient uptake [12], and also to counteract pathogens and plant diseases [13,14].

For the tomato (*Solanum lycopersicum*), grafting is also seen as a promising technique to reduce water requirement [15]. The root system of grafted plants is reported to be stronger and more efficient in absorbing water and nutrients [16], which may indirectly improve yield and fruit quality [17]. As tomato production in Germany alone reached 108,000 tons in 2019 and tends to increase worldwide [18], the water and nutrient supply must be optimized and become more sustainable. In this regard, it is useful to resort to certain rootstock cultivars that are more stress-tolerant and to improve fruit quality. Grafting was shown to increase drought-stress tolerance in tomatoes, and drought-stress tolerance varies with the rootstock used [19]. The rootstock cultivar 'Beaufort' was proven to mitigate salt stress to some extent [20], and the water- and nitrogen-use efficiency of field-grown tomato 'Florida 47' was increased when grafted onto the rootstock 'Beaufort' [21]. Moreover, the uptake of macronutrients such as nitrogen, phosphorus, and calcium was significantly improved by grafting on 'Beaufort' [22–24]. Although 'Beaufort' is proving to be an efficient rootstock, newly developed cultivars are being hybridized as rootstock that may eventually exceed this performance. For example, novel rootstock cultivars have been obtained by crossing with the wild tomato species *Solanum pennellii* accession LA716, which is considered to be more tolerant in stress conditions than related tomato species *S. lycopersicum* [25].

In this study, we compare the yield and water- and nutrient-use efficiency of two commercial tomato lines grafted onto two novel interspecific hybrid rootstocks and the commercial rootstock 'Beaufort' in two fertigation regimes. Given the drought tolerance of *S. pennellii* described in the literature, we hypothesize that *S. pennellii* × *S. lycopersicum* hybrid rootstocks can limit the yield losses of commercial tomatoes under drought to a minimum. In order to check possible influences on fruit quality, the sugar content of the fruit (°Brix), as well as flavonol and lycopene contents are also analyzed.

2. Materials and Methods

2.1. Plant Material and Treatment

The research was conducted in a greenhouse located in Meckenheim, Germany (latitude: 50°37'27.7″ N, longitude: 6°59'20.1″ E) in 2020. The two tomato (*S. lycopersicum*) F1 hybrids 'Lyterno RZ F1' (truss tomato) and the 'Tastery RZ F1' (cherry tomato) were self-grafted and grafted onto rootstock cultivars 'HUJ1', 'HUJ2' and 'Beaufort' (De-Ruiter, Bergshenhoek, NL), respectively. 'HUJ1' and 'HUJ2' are interspecific hybrids of an *S. lycopersicum* variety and the stress-tolerant wild tomato species *S. pennellii*, accession LA716 (described here [26]). In March 2020, one week prior to the scions, rootstock seeds were sown in rockwool substrate blocks (Grodan B.V.). After three weeks, right before grafting, rootstock seedlings were transferred to new rockwool blocks of size 8 cm × 8 cm × 10 cm. The grafting position was below the cotyledons of the rootstock seedlings and sealed with an appropriate clip.

Directly following this process grafted plants were put in a healing chamber for 72 h and covered with plastic bags to reduce light intensity and increase humidity to enhance the survival rate and then gradually accustomed to lower humidity levels. Thereafter, on May 10th, two plants of the same rootstock–scion combination each were planted in pots filled with *Miscanthus* × *giganteus* substrate. Twelve plants from each of the eight rootstock–scion combinations were prepared for the control and treated groups, respectively. The pots were placed in the greenhouse with an average temperature of

21.3 °C (±3.9 °C) and relative humidity of 77.9% (±16.2%). Fertigation was automatically supplied via drip irrigation. Until the third week in the greenhouse, the plants received a full supply of water and nutrients based on Groher et al. [27], to ensure the consistent growth of all plants. Subsequently, control plants received about 4 L of fertigation per day (589 L in 141 days), keeping the substrate at around 80% humidity by using a soil moisture sensor (SM150 kit, Delta T Devices Ltd., Cambridge, UK), while the treated group received about 2 L of fertigation per day (282 L in 141 days) (Figure 1). The reduced fertigation dose of technically 47.8% for treated plants is referred to as '50% reduction' from here on.

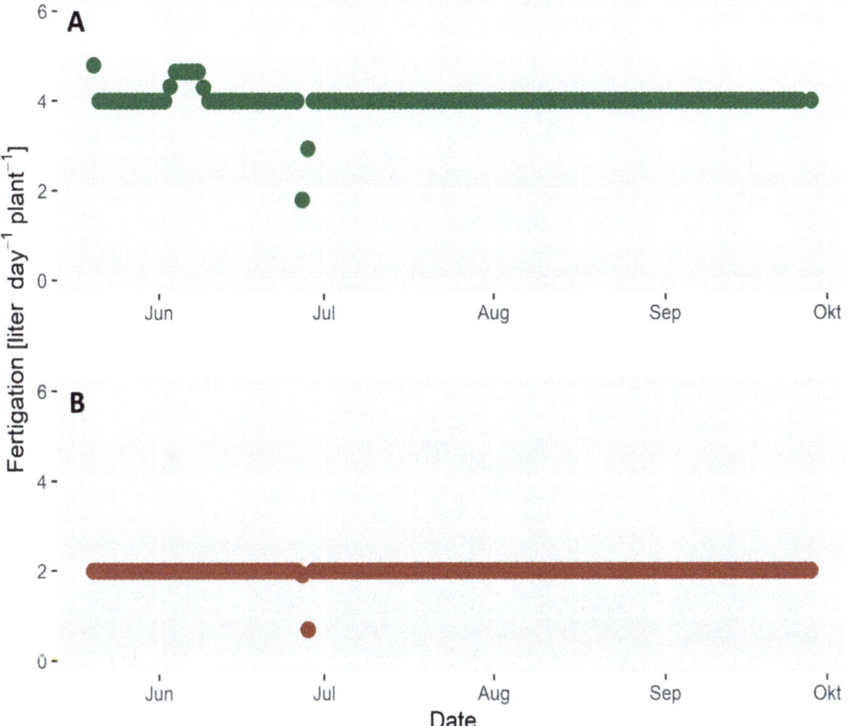

Figure 1. Daily amounts of fertigation for control (**A**) and drought stress (**B**). Occasional deviations from the desired 50% difference in fertigation between treatments occurred due to maintenance work.

2.2. Determination of Total Yield per Plant, Single Fruit Weight and Water-Use Efficiency

From the 20th week after sowing, ripe tomatoes were harvested once a week to determine the yield per plant and fruit quality in the greenhouse. Weekly harvesting was carried out until the 27th week. Harvested fruits per plant were categorized as marketable and nonmarketable fruits, based on visual assessment. The fruit number and weight of the harvested fruits of each plant were determined with a technical balance. Water-use efficiency (WUE) was calculated as the quotient of total yield per plant and fertigation per plant throughout greenhouse cultivation.

2.3. Determination of Total Soluble Solids

Total soluble solids (TTS) determination was performed on two ripe, marketable 'Lyterno' fruits (truss tomato) and five ripe, marketable 'Tastery' fruits (cocktail tomato), in order to achieve comparable amount of bulk material. Fully ripe fruits were selected from at least two and up to six plants of all rootstock−scion combinations of the control and treatment group. A blender was used to mash the tomatoes to a homogeneous mass. The type PAL^{-1} refractometer (ATAGO CO., LTD., Tokyo, Japan) was used to determine the TSS in °Brix [%] of the tomato puree.

2.4. Determination of Flavonoid and Lycopene Content

For posterior analysis of lycopene and β-carotene, we used a protocol that had already been tested before [28]. In short, tomato puree was frozen, freeze-dried and ground for 1 min using a ball mill (MM200, Retsch GmbH, Haan, Germany). Freeze-dried and ground material (0.1 g) was homogenized and extracted with a 2 mL acetone–hexane mixture (4:6). The absorbance of the extracts was measured at 453 nm, 505 nm, 645 nm and 663 nm with a UV–visible spectrophotometer (Lambda 35 UV/VIS, spectrophotometer, Perkin Elmer, Waltham, MA, USA). Concentrations of pigments were calculated according to the following equations:

$$lycopene[\text{mg}/100\text{mL}] = 0.0458_{A663} + 0.204_{A645} + 0.372_{A505} - 0.0806_{A453}$$

$$\beta - carotene[\text{mg}/100\text{mL}] = 0.216_{A663} - 0.304_{A505} + 0.452_{A453}$$

The total flavonoid content was determined according to the slightly modified methodology described by Groher et al. [27]. Briefly, a methanolic extract (80% methanol plus 1.0% hydrochloric acid) was prepared from 0.05 g freeze-dried sample material. From an aliquot of the methanolic extract, 1 mL was mixed with 0.1 mL of 0.72 mol L^{-1} sodium nitrite solution, 0.1 mL of 0.75 mol L^{-1} aluminum chloride solution, and 0.1 mL of 1 mol L^{-1} sodium hydroxide at defined time intervals. After vortexing for 10 s, 1.7 mL water was added to adjust the sample to a final volume of 3.0 mL After 30 min of incubation, absorbance was measured at 510 nm using a UV–visible spectrophotometer (Lambda 35 UV/VIS, spectrophotometer, Perkin Elmer, USA). Quercetin was used as the standard for a linear curve between 0.1 and 1.0 mg mL^{-1}, and the results were expressed as milligrams of quercetin equivalents in dry mass. Analyses were performed in triplicate, and the following equation was used to calculate flavonoid content:

$$Quercetin equivalents = 4.5709_{A510} - 0.06566$$

2.5. Data Analysis

As a preliminary step, the raw data was aggregated by the single fruit yield and total yield per plant and harvest date. For further analysis, only marketable fruits were taken into account. From those, the single fruit weight and total yield per plant were calculated. Data analysis and visualization was performed in R [29], figures were created using the R-package ggplot2 [30]. Tukey's HSD post hoc tests were used to identify differences in rootstock performance.

3. Results

As a result of the differences in fruit shape, Lyterno yields are generally higher than Tastery yields. Mean Lyterno yield per plant in optimal conditions in the seven weeks experimental period was 4906 g (±937 g), with no significant influence of the rootstock used (Figure 2A). For Tastery, the mean yield per plant was 2603 g (±250 g) (Figure 2C). When fertigation was reduced, Lyterno yields dropped by around 39% on average to 2961 g (±899 g) (Figure 2B), while the mean reduction in Tastery yields (24%) was a rather moderate 1978 g (±388 g) (Figure 2D). WUE is calculated as the quotient of yield per plant and fertigation per plant. The latter was held constant across rootstock–scion combinations within treated and control plants. The generally observed relative increase in WUE of stressed plants was higher in Tastery than in Lyterno (Tastery: 4.42 to 7.01 g/L; Lyterno: 8.33 to 10.50 g/L; Figure 2ii).

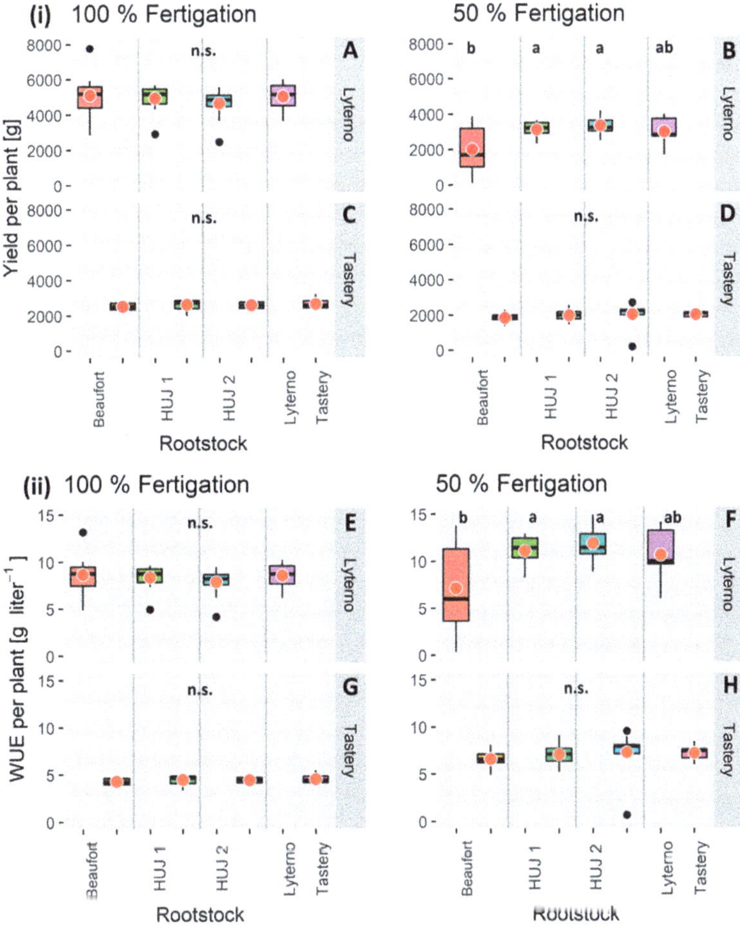

Figure 2. Average yield ((**i**), top) and water-use efficiency ((**ii**), bottom) per plant (n = 12) of scions from cultivar 'Lyterno' (**A,B,E,F**) and 'Tastery' (**C,D,G,H**) grafted onto different rootstocks and supplied with optimal (**A,C,E,G**) and reduced (**B,D,F,H**) fertigation. n.s.: not significant, *p*-value 0.05.

The large standard deviations in the Lyterno yields under reduced fertigation can partly be explained through the rootstock effect: the results of an analysis of variance and subsequent Tukey's HSD tests are presented in Table S1. Plants grafted onto Beaufort yields were significantly lower than those grafted onto HUJ 1 and HUJ 2. Self-grafted controls tended to yield lower than HUJ 1 and HUJ 2, but differences were not significant.

The results of fruit quality analyses of scion 'Lyterno' are presented in Figures 3 and 4. As no change in yield of scion 'Tastery' related to the different rootstock was found (Figure 2C,D), our further analyses focus on Lyterno. Tastery fruit quality parameters are presented in supplementary Figures S1 and S2.

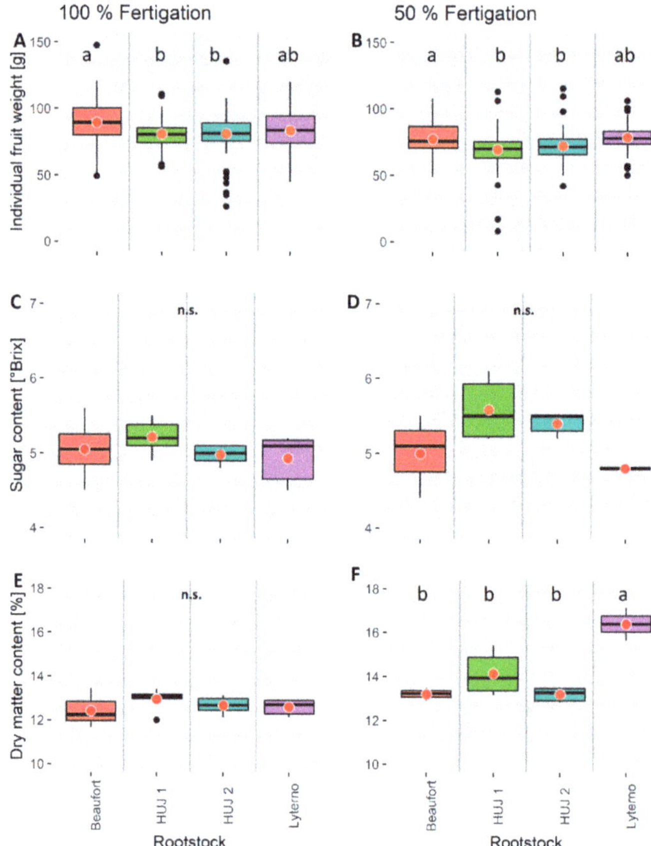

Figure 3. Quality parameters of 'Lyterno' fruits grafted onto four different rootstocks and supplied with optimal (**A**,**C**,**E**) and reduced (**B**,**D**,**F**) fertigation: Individual fruit weight (g) (n = 49–84), (**A**,**B**); total soluble solids ('sugar') content of ripe fruits (°Brix) (fruits of n = 2–6 plants), (**C**,**D**); fruit dry matter content [%]; (n = 2–6) (**E**,**F**). n.s.: not significant, *p*-value 0.05.

The individual weight of the Lyterno fruits drops by 11.2% from 83.1 g (±15.0 g) to 73.8 g (±13.1 g) when fertigation is reduced. Individual fruit weight of control plants grafted onto Beaufort is significantly higher than grafted onto both HUJ 1 and HUJ 2 (*p* < 0.01) (Figure 3A). In the reduced fertigation regime, self-grafted plants produce on average heavier fruits than scions grafted onto HUJ 1 and HUJ 2 (Figure 3B).

The pattern changes when looking at fruit sugar content. Sugar contents [°Brix] are generally slightly increased when fertigation is reduced: From 5.05 °Brix (±0.29 °Brix) to 5.32 °Brix (±0.43 °Brix). Probably as a result of few marketable fruits from stressed plants grafted onto different rootstocks being available for laboratory analyses (n = 2–6), no significant differences in sugar content can be attributed to the different rootstocks. However, a tendency to higher sugar contents in stressed fruits grafted onto HUJ 1 and HUJ 2 is depicted in Figure 3D.

Under the influence of drought stress, the average dry matter content of Lyterno fruits increased from 12.66% (±0.51%) to 13.94% (±1.23%) (Figure 3E,F). While there were no rootstock dependent differences under optimal fertigation application, the dry matter content of self-grafted plants was higher under reduced fertigation application.

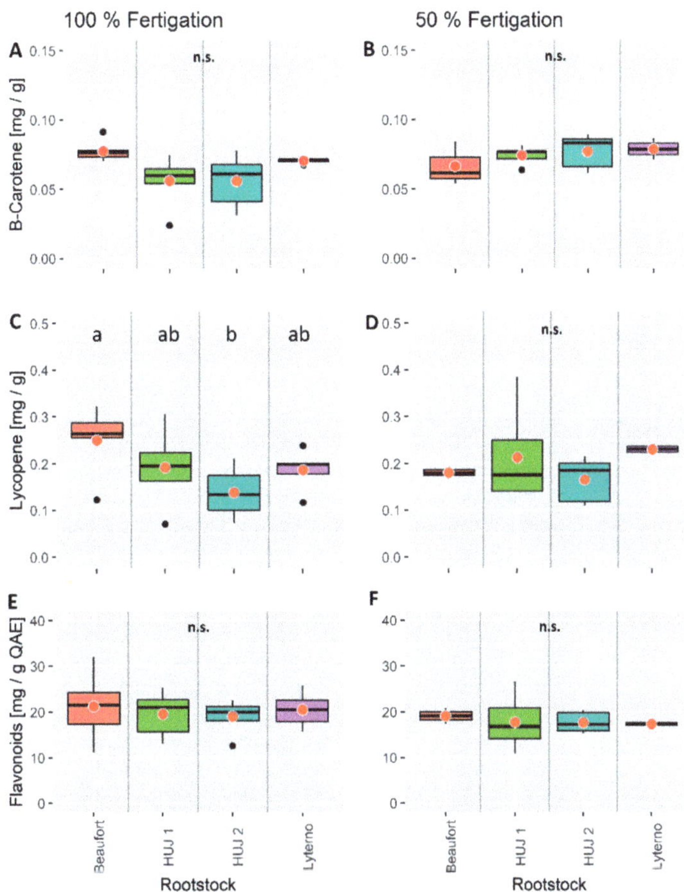

Figure 4. Quality parameters of 'Lyterno' fruits grafted onto four different rootstocks and supplied with optimal (**A,C,E**) and reduced (**B,D,F**) fertigation: β-carotene content (mg/g DM), (**A,B**); lycopene content (mg/g DM), (**C,D**); flavonoid content (mg/g DM quercetin equivalents) (**E,F**); n = (2–6). n.s.: not significant, *p*-value 0.05.

Only minor differences in the β-carotene, lycopene and flavonoid contents of the 'Lyterno' fruits grafted onto different rootstocks and supplied with optimal or reduced fertigation were detected (Figure 4).

As contents are presented in milligram per gram dry weight (Figure 4), and the dry matter of stressed fruits was on average higher than for control fruits (Figure 3E,F), the contents of β-carotene, lycopene and flavonoids per fresh matter were higher for stressed than for control fruits (not shown).

A comparison of the lycopene and β-carotene contents of the fruits indicates a higher responsiveness of β-carotene than lycopene to the applied changes in fertigation.

4. Discussion

A major objective of the present study was to exploit the potential of new genetic material for tomato grafting to target the necessary improvements for horticulture in the scope of climate-related challenges in future. We, therefore, verified to what extent novel hybrid rootstocks (*S. pennellii* × *S. lycopersicum*) could minimize tomato yield loss under limited irrigation conditions, as scarcity of water for irrigation will increase in the future [31]. Several studies have shown that remedial action can be provided by combining resilient rootstock and high yielding scions, while the physiological exchange of these plant parts significantly influences overall performance and resilience to environmental stressors. In addition, grafting can improve vegetable quality by enhancing biosynthesis of endogenous phytohormones, and by acquiring and transporting mineral nutrients [32,33]. Fruit flavor and nutritional value can be furthermore positively influenced by altered secondary metabolites, and by the concentration of health-promoting compounds, such as lycopene, carotenoids and amino acids [33,34]. Congruently, tomatoes grafted onto efficient rootstocks are also both nutritionally promising and at the same time more tolerant to abiotic and biotic stressors [35], which, besides the sustainable use of resources, in turn offers additional advantages in the case of more frequent and severe incidence of climate-related diseases.

4.1. Impact of Different Rootstocks on Scion Performance

Based on plant performance in yield and water-use efficiency, given the data presented (Figure 2i), we can conclude that the two novel rootstocks tested were indeed limiting yield losses of the truss-tomato cultivar Lyterno under drought stress conditions, as compared to the commercial rootstock 'Beaufort'. At the same time, underlining the potential of the new rootstock cultivars in optimal conditions, the yields of scions grafted onto the novel rootstocks were not different from those grafted onto 'Beaufort'.

While for the cocktail tomato Tastery, yields did not change depending on the rootstock used, either in optimal or drought conditions, where 5–10% losses were recorded.

The identical fertigation for plants with both scions apparently resulted in Lyterno being more stressed than Tastery, as can be seen from the greater yield reduction (Figure 2) and the more pronounced differences in fruit quality (Figure 3) for Lyterno.

This is in line with previous studies showing high-yielding performance under optimal conditions when grafted onto 'Beaufort' being advantageous for several scion cultivars [23,36]. Here, 'Beaufort' was ascertained as positive in greater stem height and diameter, as well as epigeous dry biomass [23]. Under stress conditions, for example, saline and drought treatment, 'Beaufort' was shown to be more tolerant than other genotypes, such as 'Body', 'Heman', 'Resistar', 'Spirit', 'Vigomax' and 'Vedi' [20], and 'Maxifort', 'Unifort', 'Vedi', 'Kemerit', 'King' 'Kong', 'Spirit', 'Resistar', '500292' and 'Toro' [19], respectively.

With a similarly promising performance, the novel rootstocks 'HUJ1' and 'HUJ2' that are based on the *S. pennellii* introgression lines (ILs), are considered to maintain stable growth when exposed to salt stress [25]. When stressed, the ILs performed significantly better at maintaining plant growth than the salt-tolerant wild species, as well as parental line LA716.

Nevertheless, for commercially relevant scions, we focused on two production varieties commonly used in central Europe and North America, namely the cocktail tomato 'Tastery' and the truss-tomato 'Lyterno' [37].

At this stage, it remains an open question whether Tastery is generally less responsive to different rootstocks, or whether the lack of rootstock-related differences is due to the relatively mild stress. The fact that yield and fruit weights were lower overall under drought stress was to be expected and was in line with other studies [38,39]. A surprising result was the consistently good performance of the self-grafted plants in our experiment, contradictory to other studies of different cultivars and scion–rootstock combinations [40,41]. The comparably good performance of our 'Tastery–Tastery' plants might be referred to the internal physiological communication between rootstock and scion, being identical for self-grafted plants. This essentially coordinated exchange of endogenous phytohormones, mineral elements, mRNA, and proteins [42], as well as relevant transporters (either influx or efflux carriers) might affect a smoother long-distance signaling of phytohormones through the plant vascular system [43] of genetically identical plants compared to different scion and rootstock genotypes.

4.2. Effects of Rootstocks on Water-Use Efficiency during Cultivation Cycle

In the scope of future challenges of water shortage, water-use efficiency (WUE) is a relevant option to determine plant performance. Reduced irrigation either by an improved WUE of (grafted) tomato genotypes [44] or by effective irrigation management during fruit developmental stages ensures sufficient yield and fruit quality [38].

However, under artificial conditions, the data on yield and WUE should be viewed with caution. Indeed, in order to assess validated data, the irrigation demands of every single plant would need to be considered. On the other hand, such an approach is not yet feasible for commercial-like conditions. Highly sophisticated greenhouse systems based on artificial intelligence might improve the current situation in the near future [45]. Nevertheless, in our experimental case, water demand and, thus, WUE would need to be explored under two separate irrigation regimes to adapt to the cultivar-specific needs of Tastery and Lyterno and/or rootstocks in more depth.

Furthermore, we focused on a rather short time period of productive cultivation cycles. The overall yield performance (Figure 2i) may therefore be used as an indicator for differences in yield among rootstocks, but the absolute values cannot be compared to commercial yields. While production cycles in commercial cultivation in Central European greenhouses last up to 10 months, our calculation of WUE only includes yields from a seven-week period. Moreover, we only started harvesting once all plants produced stable yields, rather than already picking the first tomatoes in the season. By the same token, WUE data presented in Figure 2ii may be used as an indicator for different WUE among rootstocks, but cannot be compared to WUE in commercial production: On the one hand, yield is not comparable to commercial production for the reasons above and on the other hand, water-use data is also different from commercial production, as, e.g., water consumption during plant germination, between grafting and planting out, and finally the water used to flood the *Miscanthus* substrate after transplanting to ensure plant stand establishment, are not included in our calculations in Figure 2ii. Since these factors were the same for all our experimental plants and rootstock−scion combinations, they can be neglected, but need to be considered when calculating the overall water use and thus, WUE of genotypes.

4.3. Rootstock Influence on Fruit Quality Parameters

Fruit quality, such as its nutritional value, taste and odor, is of great importance for consumers and therefore for producers [46]. Grafting has a significant effect on quality parameters improving health-promoting properties and flavor by the altered composition of secondary metabolites [33,34].

Looking at the characteristics of the fruits of plants grafted onto different rootstocks, some interesting results emerged. The two *S. pennellii*-based rootstocks seemed to favor the production of smaller fruits. Lyterno fruits from plants on these rootstocks were almost universally significantly lighter than the fruits on Beaufort or self-grafted Lyterno (Figure 3A,B). No significant differences were seen for Tastery, but again there is a slight trend toward smaller fruits on *S. pennellii*-based rootstocks (Figure S1A,B).

The fact that *S. pennellii* itself produces very small fruits with a fruit weight < 16 g [47] could be an indication that this trait has been transferred from the rootstock to the scion.

In contrast, we observed a fruit size-enhancing effect of 'Beaufort', resulting in the comparably high individual fruit weight of Lyterno, as was previously described for this type of vigorous interspecific rootstocks [10,11,36,48]. This effect on fruit size is significantly influenced by the grafting combination [41], and needs to be distinguished from yield performance that can also be attributed to an enhanced number of tomato fruits rather than an increased average fruit weight [49].

It is known that rootstock mediation affects fruit quality, due to altered water and nutrient uptake, and changes in the source−sink balance impacting ripening behavior [10,16]. Therefore, it can be concluded that some rootstocks favor early fruit set, as well as rapid fruit ripening [50]. In particular, the possible faster fruit ripening could have resulted in the lower fruit weight of the Lyterno grafted onto 'HUJ1' and 'HUJ2'. Nevertheless, individual fruit weight was further reduced under 50% fertigation, implying that drought could also have accelerated tomato ripening [27].

Both overall increased dry matter and sugar content, and reduced fruit weight are a logical result of drought stress: with optimal water availability in the root zone, the high sugar concentration in the fruit ensures that water is transported from the root zone into the fruit by osmotic effects. When there is a lack of water, this 'sucking in' of water is not possible, and there is an increased content of soluble substances with a simultaneously reduced fruit weight [51].

The opposite trends observed in Lyterno for yield and fruit weight of *S. pennellii*-based rootstocks are noteworthy: total yield was relatively high when these rootstocks were used under drought stress (Figure 2B), but at the same time individual fruit weight was relatively low (Figure 3B and Figure S1). This can be explained by the fact that more but smaller fruits reached maturity [49].

These fruits, then, also tend to be sweeter (Figure 3D), which can be seen as another potential advantage provided by the novel rootstocks. Concerning taste and quality, this might be an approach to target current consumer needs.

Surprisingly, in regard to secondary metabolites in Lyterno fruits (Figure 4), we only found a significantly reduced lycopene content for the rootstock 'HUJ2' under optimal fertigation conditions. Here, 'Beaufort' showed the overall highest contents of β-carotene, lycopene and flavonoids. This is in contrast to previous findings, where 'Beaufort' resulted in a decrease of lycopene concentration of the scions 'Jeremy' and 'Jack' [52,53]. This might be attributed to the specific interaction of rootstock and scion [10,11]. However, under 50% fertigation, slightly reduced but comparably high metabolite contents were analyzed for Lyterno, when individual fruit weight and yield decreased. This is in agreement with previous field trials, where soluble solids, lycopene, and sugar content were increased in the tomato fruit, while fruit yield decreased between 1.9%–18.2% [54]. A similar increase in tomato fruit quality parameters was shown for a pot experiment with a two-thirds yield decrease under irrigation with 0.3% salt concentration in soil [55]. Based on our experiment, it could be argued that the reduction in fertigation also resulted in a reduced volume of 'Lyterno' tomatoes and thus both sugar content and metabolite contents increased in the fruit.

4.4. Future Perspective of Renewable Substrate for Hydroponics

Although there were no distinct patterns in the individual grafting combinations with lower fertigation dosages in our experiment, this project is still a sustainable approach for food production targeting resource-efficient tomato production by adjusting both plant performance (via grafting) and crop management. Eventually, a sustainable and more environmentally friendly substrate of *Miscanthus* × *giganteus* was used. In the study of Kraska et al. [56], it was shown that *Miscanthus* as shreds, chips and fibers does not differ much in comparison to rockwool as a substrate for the cumulative fruit yield (g/plant), which highlights in our experiment the fact, that the use of *Miscanthus* × *giganteus* should not negatively impact the yield measurement. Moreover, the yields and fruit quality of cucumber and tomatoes grown on *Miscanthus* as a potential alternative to rockwool as a growing medium for vegetables [57,58] and is overall seen as a promising alternative to rockwool as a growing medium for tomatoes in soilless cultivation [56].

To determine the optimal water and nutrient requirements in grafted 'Lyterno' and 'Tastery' tomato plants, further studies will need to be conducted in order to gain a more detailed insight into the interaction between rootstock and scions, as well as the potential to identify the genotype combinations being more tolerant towards drought stress and nutrient limitations. Ultimately, besides sustainability, it also makes financial sense to graft plants and thereby apply less water or nutrients for the same yield, than not grafting and thereby use up many limited resources, without losing fruit quality.

Supplementary Materials: The following are available online at https://www.mdpi.com/article/10.3390/horticulturae7100355/s1, Table S1: Tukey comparison of mean differences for plants with the scion Lyterno under stress conditions. Figure S1: Quality parameters of "Tastery" fruits grafted on four different rootstocks and supplied with optimal and reduced fertigation. Figure S2: Quality parameters of "Tastery" fruits grafted on four different rootstocks and supplied with optimal and reduced fertigation. Figure S3: Comparison of lycopene and β-carotene contents of ripe tomatoes from 8 rootstock * scion – combinations supplied with optimal and reduced fertigation.

Author Contributions: Conceptualization, J.E., S.R.-S.; data analysis, J.E., A.B., P.B.; methodology, J.E.; writing—original draft preparation, J.E., A.B., P.B.; writing—review and editing, J.E., S.R.-S., A.B. and P.B. All authors have read and agreed to the published version of the manuscript.

Funding: J.E. received funding from the European Union's Horizon 2020 research and innovation program under the Grant Agreement No. (727929) (TOMRES).

Institutional Review Board Statement: Not applicable.

Informed Consent Statement: Not applicable.

Data Availability Statement: Raw data is available on request from the corresponding author.

Acknowledgments: The authors thank Dani Zamir, Hebrew University of Jerusalem, for providing *S. pennellii* × *S. lycopersicum* seeds. We also thank the technical staff of the Institute for Crop Science and Resource Conservation of Bonn University for their cooperation.

Conflicts of Interest: The authors declare no conflict of interest.

References

1. Sánchez-Rodríguez, E.; Leyva, R.; Constán-Aguilar, C.; Romero, L.; Ruiz, J.M. Grafting under Water Stress in Tomato Cherry: Improving the Fruit Yield and Quality. *Ann. Appl. Biol.* **2012**, *161*, 302–312. [CrossRef]
2. Ullah, H.; Santiago-Arenas, R.; Ferdous, Z.; Attia, A.; Datta, A. Chapter Two—Improving Water Use Efficiency, Nitrogen Use Efficiency, and Radiation Use Efficiency in Field Crops under Drought Stress: A Review. *Adv. Agron.* **2019**, *156*, 109–157. [CrossRef]
3. Cherry, C.; Scott, C.; Barrett, J.; Pidgeon, N. Public Acceptance of Resource-Efficiency Strategies to Mitigate Climate Change. *Nat. Clim. Chang.* **2018**, *8*, 1007–1012. [CrossRef]
4. Cuartero, J.; Bolarin, M.C.; Asins, M.J.; Moreno, V. Increasing Salt Tolerance in the Tomato. *J. Exp. Bot.* **2006**, *57*, 1045–1058. [CrossRef] [PubMed]
5. Dempewolf, H.; Baute, G.; Anderson, J.; Kilian, B.; Smith, C.; Guarino, L. Past and Future Use of Wild Relatives in Crop Breeding. *Crop. Sci.* **2017**, *57*, 1070–1082. [CrossRef]
6. Bolger, A.; Scossa, F.; Bolger, M.E.; Lanz, C.; Maumus, F.; Tohge, T.; Quesneville, H.; Alseekh, S.; Sørensen, I.; Lichtenstein, G.; et al. The Genome of the Stress Tolerant Wild Tomato Species Solanum Pennellii. *Nat. Genet.* **2014**, *46*, 1034–1038. [CrossRef]
7. Zsögön, A.; Čermák, T.; Naves, E.R.; Notini, M.M.; Edel, K.H.; Weinl, S.; Freschi, L.; Voytas, D.F.; Kudla, J.; Peres, L.E.P. De Novo Domestication of Wild Tomato Using Genome Editing. *Nat. Biotechnol.* **2018**, *36*, 1211–1216. [CrossRef]
8. Zimny, T.; Sowa, S.; Tyczewska, A.; Twardowski, T. Certain New Plant Breeding Techniques and Their Marketability in the Context of EU GMO Legislation—Recent Developments. *New Biotechnol.* **2019**, *51*, 49–56. [CrossRef] [PubMed]
9. Callaway, E. CRISPR Plants Now Subject to Tough GM Laws in European Union. *Nature* **2018**, *560*, 16. [CrossRef]
10. Kyriacou, M.C.; Rouphael, Y.; Colla, G.; Zrenner, R.; Schwarz, D. Vegetable Grafting: The Implications of a Growing Agronomic Imperative for Vegetable Fruit Quality and Nutritive Value. *Front. Plant Sci.* **2017**, *8*, 741. [CrossRef]
11. Rouphael, Y.; Schwarz, D.; Krumbein, A.; Colla, G. Impact of Grafting on Product Quality of Fruit Vegetables. *Sci. Hortic.* **2010**, *127*, 172–179. [CrossRef]
12. Alan, O.; Ozdemir, N.; Gunen, Y. Effect of Grafting on Watermelon Plant Growth, Yield and Quality. *J. Agron.* **2007**, *6*, 362–365. [CrossRef]
13. López-Pérez, J.-A.; Le Strange, M.; Kaloshian, I.; Ploeg, A.T. Differential Response of Mi Gene-Resistant Tomato Rootstocks to Root-Knot Nematodes (Meloidogyne Incognita). *Crop. Prot.* **2006**, *25*, 382–388. [CrossRef]
14. Rivard, C.L.; O'Connell, S.; Peet, M.M.; Welker, R.M.; Louws, F.J. Grafting Tomato to Manage Bacterial Wilt Caused by *Ralstonia Solanacearum* in the Southeastern United States. *Plant Dis.* **2012**, *96*, 973–978. [CrossRef] [PubMed]
15. Estan, M.T.; Martinez-Rodriguez, M.M.; Perez-Alfocea, F.; Flowers, T.J.; Bolarin, M.C. Grafting Raises the Salt Tolerance of Tomato through Limiting the Transport of Sodium and Chloride to the Shoot. *J. Exp. Bot.* **2005**, *56*, 703–712. [CrossRef] [PubMed]
16. Savvas, D.; Colla, G.; Rouphael, Y.; Schwarz, D. Amelioration of Heavy Metal and Nutrient Stress in Fruit Vegetables by Grafting. *Sci. Hortic.* **2010**, *127*, 156–161. [CrossRef]

17. Flores, F.B.; Sanchez-Bel, P.; Estañ, M.T.; Martinez-Rodriguez, M.M.; Moyano, E.; Morales, B.; Campos, J.F.; Garcia-Abellán, J.O.; Egea, M.I.; Fernández-Garcia, N.; et al. The Effectiveness of Grafting to Improve Tomato Fruit Quality. *Sci. Hortic.* **2010**, *125*, 211–217. [CrossRef]
18. Statistisches-Bundesamt. *Erntemenge von Tomaten in Deutschland in Den. Jahren 2001 Bis 2019 (in 1.000 Tonnen)*; Statistisches Bundesamt Destatis: Wiesbaden, Germany, 2020.
19. Altunlu, H.; Gul, A. Increasing Drought Tolerance of Tomato Plants by Grafting. *Acta Hortic.* **2012**, 183–190. [CrossRef]
20. Öztekin, G.B.; Tuzel, Y. Salinity Response of Some Tomato Rootstocks at Seedling Stage. *Afr. J. Agric. Res.* **2011**, *6*, 4726–4735. [CrossRef]
21. Djidonou, D.; Zhao, X.; Simonne, E.H.; Koch, K.E.; Erickson, J.E. Yield, Water-, and Nitrogen-Use Efficiency in Field-Grown, Grafted Tomatoes. *HortScience* **2013**, *48*, 485–492. [CrossRef]
22. Ruiz, J.M.; Romero, L. Nitrogen Efficiency and Metabolism in Grafted Melon Plants. *Sci. Hortic.* **1999**, *81*, 113–123. [CrossRef]
23. Leonardi, C.; Giuffrida, F. Variation of Plant Growth and Macronutrient Uptake in Grafted Tomatoes and Eggplants on Three Different Rootstocks. *Eur. J. Hortic. Sci.* **2006**, *71*, 97–101.
24. Goto, R.; de Miguel, A.; Marsal, J.I.; Gorbe, E.; Calatayud, A. Effect of Different Rootstocks on Growth, Chlorophyll a Fluorescence and Mineral Composition of Two Grafted Scions of Tomato. *J. Plant Nutr.* **2013**, *36*, 825–835. [CrossRef]
25. Frary, A.; Keleş, D.; Pinar, H.; Göl, D.; Doğanlar, S. NaCl Tolerance in Lycopersicon Pennellii Introgression Lines: QTL Related to Physiological Responses. *Biol. Plant.* **2011**, *55*, 461–468. [CrossRef]
26. Ofner, I.; Lashbrooke, J.; Pleban, T.; Aharoni, A.; Zamir, D. Solanum Pennellii Backcross Inbred Lines (BILs) Link Small Genomic Bins with Tomato Traits. *Plant J.* **2016**, *87*, 151–160. [CrossRef] [PubMed]
27. Groher, T.; Schmittgen, S.; Fiebig, A.; Noga, G.; Hunsche, M. Suitability of Fluorescence Indices for the Estimation of Fruit Maturity Compounds in Tomato Fruits. *J. Sci. Food Agric.* **2018**, *98*, 5656–5665. [CrossRef] [PubMed]
28. Nagata, M.; Yamashita, I. Simple Method for Simultaneous Determination of Chlorophyll and Carotenoids in Tomato Fruit. *Jpn. Soc. Food Sci. Technol.* **1992**, *39*, 925–928. [CrossRef]
29. R-Core-Team. *R: A Language and Environment for Statistical Computing*; R Foundation for Statistical Computing: Vienna, Austria, 2020.
30. Wickham, H. ggplot2. In *Ggplot2: Elegant Graphics for Data Analysis*; Springer: New York, NY, USA, 2009; ISBN 978-0-387-98141-3.
31. Mancosu, N.; Snyder, R.L.; Kyriakakis, G.; Spano, D. Water Scarcity and Future Challenges for Food Production. *Water* **2015**, *7*, 975–992. [CrossRef]
32. Wang, Q.; Men, L.; Gao, L.; Tian, Y. Effect of Grafting and Gypsum Application on Cucumber (*Cucumis Sativus* L.) Growth under Saline Water Irrigation. *Agric. Water Manag.* **2017**, *188*, 79–90. [CrossRef]
33. Lee, J.-M.; Kubota, C.; Tsao, S.J.; Bie, Z.; Echevarria, P.H.; Morra, L.; Oda, M. Current Status of Vegetable Grafting: Diffusion, Grafting Techniques, Automation. *Sci. Hortic.* **2010**, *127*, 93–105. [CrossRef]
34. Davis, A.R.; Perkins-Veazie, P.; Hassell, R.; Levi, A.; King, S.R.; Zhang, X. Grafting Effects on Vegetable Quality. *Am. Soc. Hortic. Sci.* **2021**, *43*, 1670–1672. [CrossRef]
35. Singh, H.; Kumar, P.; Chaudhari, S.; Edelstein, M. Tomato Grafting: A Global Perspective. *HortScience* **2017**, *52*, 1328–1336. [CrossRef]
36. Turhan, A.; Ozmen, N.; Serbeci, M.S.; Seniz, V. Effects of Grafting on Different Rootstocks on Tomato Fruit Yield and Quality. *Hortic. Sci.* **2011**, *38*, 142–149. [CrossRef]
37. Rijk-Zwaan. *Rijk Zwaan North. America | Product Catalogue Fruitcrops*; Rijk Zwaan Canada Leamington Demonstration House: Leamington, ON, Canada, 2020.
38. Cui, J.; Shao, G.; Lu, J.; Keabetswe, L.; Hoogenboom, G. Yield, Quality and Drought Sensitivity of Tomato to Water Deficit during Different Growth Stages. *Sci. Agric.* **2020**, *77*, 1–9. [CrossRef]
39. Abdulaziz, A.-H.; Abdulrasoul, A.-O.; Thabit, A.; Hesham, A.-R.; Khadejah, A.; Saad, M.; Abdullah, O. Tomato Grafting Impacts on Yield and Fruit Quality under Water Stress Conditions. *J. Exp. Biol. Agric. Sci.* **2017**, *5*, 136–147. [CrossRef]
40. Khah, E.M.; Kakava, E.; Mavromatis, A.; Chachalis, D.; Goulas, C. Effect of Grafting on Growth and Yield of Tomato (Lycopersicon Esculentum Mill.) in Greenhouse and Open-Tield. *J. Appl. Hortic.* **2006**, *8*, 3–7. [CrossRef]
41. Schwarz, D.; Öztekin, G.B.; Tüzel, Y.; Brückner, B.; Krumbein, A. Rootstocks Can Enhance Tomato Growth and Quality Characteristics at Low Potassium Supply. *Sci. Hortic.* **2013**, *149*, 70–79. [CrossRef]
42. Lu, X.; Liu, W.; Wang, T.; Zhang, J.; Li, X.; Zhang, W. Systemic Long-Distance Signaling and Communication between Rootstock and Scion in Grafted Vegetables. *Front. Plant Sci.* **2020**, *11*, 460. [CrossRef]
43. Lacombe, B.; Achard, P. Long-Distance Transport of Phytohormones through the Plant Vascular System. *Curr. Opin. Plant Biol.* **2016**, *34*, 1–8. [CrossRef]
44. Kumar, P.; Rouphael, Y.; Cardarelli, M.; Colla, G. Vegetable Grafting as a Tool to Improve Drought Resistance and Water Use Efficiency. *Front. Plant Sci.* **2017**, *8*, 1130. [CrossRef]
45. Hemming, S.; de Zwart, F.; Elings, A.; Petropoulou, A.; Righini, I. Cherry Tomato Production in Intelligent Greenhouses—Sensors and AI for Control of Climate, Irrigation, Crop Yield, and Quality. *Sensors* **2020**, *20*, 6430. [CrossRef] [PubMed]
46. Liu, Z.; Alseekh, S.; Brotman, Y.; Zheng, Y.; Fei, Z.; Tieman, D.M.; Giovannoni, J.J.; Fernie, A.R.; Klee, H.J. Identification of a Solanum Pennellii Chromosome 4 Fruit Flavor and Nutritional Quality-Associated Metabolite QTL. *Front. Plant Sci.* **2016**, *7*. [CrossRef] [PubMed]

47. Talekar, N.S.; Opeña, R.T.; Hanson, P. Helicoverpa Armigera Management: A Review of AVRDC's Research on Host Plant Resistance in Tomato. *Crop. Prot.* **2006**, *25*, 461–467. [CrossRef]
48. Pogonyi, Á.; Pék, Z.; Helyes, L.; Lugasi, A. Effect of Grafting on the Tomato's Yield, Quality and Main Fruit Components in Spring Forcing. *Acta Aliment.* **2005**, *34*, 453–462. [CrossRef]
49. Savvas, D.; Savva, A.; Ntatsi, G.; Ropokis, A.; Karapanos, I.; Krumbein, A.; Olympios, C. Effects of Three Commercial Rootstocks on Mineral Nutrition, Fruit Yield, and Quality of Salinized Tomato. *J. Plant Nutr. Soil Sci.* **2011**, *174*, 154–162. [CrossRef]
50. Fallik, E.; Ilic, Z. Grafted Vegetables—The Influence of Rootstock and Scion on Postharvest Quality. *Folia Hortic.* **2014**, *26*, 79–90. [CrossRef]
51. Prudent, M.; Causse, M.; Génard, M.; Tripodi, P.; Grandillo, S.; Bertin, N. Genetic and Physiological Analysis of Tomato Fruit Weight and Composition: Influence of Carbon Availability on QTL Detection. *J. Exp. Bot.* **2009**, *60*, 923–937. [CrossRef]
52. M

isković, A.; Ilin, Z.; Marković, V. Effect of Different Rootstock Type on Quality and Yield of Tomato Fruits. *Acta Hortic.* **2009**, *807*, 619–624. [CrossRef]
53. Riga, P.; Benedicto, L.; García-Flores, L.; Villaño, D.; Medina, S.; Gil-Izquierdo, Á. Rootstock Effect on Serotonin and Nutritional Quality of Tomatoes Produced under Low Temperature and Light Conditions. *J. Food Compos. Anal.* **2016**, *46*, 50–59. [CrossRef]
54. Li, J.; Chen, J.; Qu, Z.; Wang, S.; He, P.; Zhang, N. Effects of Alternating Irrigation with Fresh and Saline Water on the Soil Salt, Soil Nutrients, and Yield of Tomatoes. *Water* **2019**, *11*, 1693. [CrossRef]
55. Yang, H.; Du, T.; Mao, X.; Ding, R.; Shukla, M.K. A Comprehensive Method of Evaluating the Impact of Drought and Salt Stress on Tomato Growth and Fruit Quality Based on EPIC Growth Model. *Agric. Water Manag.* **2019**, *213*, 116–127. [CrossRef]
56. Kraska, T.; Kleinschmidt, B.; Weinand, J.; Pude, R. Cascading Use of Miscanthus as Growing Substrate in Soilless Cultivation of Vegetables (Tomatoes, Cucumbers) and Subsequent Direct Combustion. *Sci. Hortic.* **2018**, *235*, 205–213. [CrossRef]
57. FAO. *Good Agricultural Practices for Greenhouse Vegetable Crops: Principles for Mediterranean Climate Areas*; FAO Plant Production and Protection Paper; Food and Agricultural Organization of the United Nations (FAO): Rome, Italy, 2013; ISBN 978-92-5-107650-7.
58. Barrett, G.E.; Alexander, P.D.; Robinson, J.S.; Bragg, N.C. Achieving Environmentally Sustainable Growing Media for Soilless Plant Cultivation Systems—A Review. *Sci. Hortic.* **2016**, *212*, 220–234. [CrossRef]

 *horticulturae*

Article

Opportunities of Reduced Nitrogen Supply for Productivity, Taste, Valuable Compounds and Storage Life of Cocktail Tomato

Lilian Schmidt [1,2,*] and Jana Zinkernagel [2]

[1] Soil Science and Plant Nutrition, Hochschule Geisenheim University, Von-Lade-Strasse 1, 65366 Geisenheim, Germany

[2] Vegetable Crops, Hochschule Geisenheim University, Von-Lade-Strasse 1, 65366 Geisenheim, Germany; Jana.Zinkernagel@hs-gm.de

* Correspondence: Lilian.Schmidt@hs-gm.de

Abstract: Vegetable production requires high nutrient input for ensuring high quality and high yield. As this is ecologically disadvantageous, it is necessary to determine if nitrogen (N) fertilization can be reduced without negative effects on productivity. For quality reasons, the effects of reduced N supply on taste, valuable compounds and storage life must be elucidated in parallel. This study examines whether reducing the N supply of cocktail tomatoes by 50% to recommendations affects the yield and quality of tomato fruits. Three varieties with different skin colors, yellow-orange ('Apresa'), red ('Delioso') and brown ('Bombonera'), were grown in soil in a greenhouse and harvested at the red-ripen stage. Quality parameters were assessed at harvest and after eight-day storage. Total yield decreased exclusively with 'Bombonera' due to reduced fruit weight. Firmness of the fruit pulp, concentrations of minerals, soluble solid contents, total acidity, total phenolics and liposoluble pigments of fruits were not influenced. However, storage affected chemical compositions positively, as shown by increased antioxidants. Descriptive sensory analyses revealed no impact of reduced N supply. From the perspective of the yield, quality and shelf life of fruits, reducing the N supply by 50% offers opportunities for the three cocktail tomato varieties in soil cultivation.

Keywords: tomato; product quality; nitrogen; shelf life; carotenoids; antioxidants; taste

Citation: Schmidt, L.; Zinkernagel, J. Opportunities of Reduced Nitrogen Supply for Productivity, Taste, Valuable Compounds and Storage Life of Cocktail Tomato. *Horticulturae* **2021**, *7*, 48. https://doi.org/10.3390/horticulturae7030048

Academic Editors: Wilfried Rozhon and Antonio Ferrante

Received: 29 January 2021
Accepted: 8 March 2021
Published: 11 March 2021

Publisher's Note: MDPI stays neutral with regard to jurisdictional claims in published maps and institutional affiliations.

Copyright: © 2021 by the authors. Licensee MDPI, Basel, Switzerland. This article is an open access article distributed under the terms and conditions of the Creative Commons Attribution (CC BY) license (https://creativecommons.org/licenses/by/4.0/).

1. Introduction

Nitrogen (N) is the most important mineral element for plants and, as a constituent of amino acids and proteins, often limiting for their growth processes [1]. In the context of plant production, N influences the appearance and internal quality of plant products, their yield and their shelf life. The supply of N is therefore decisive for the quality of food crops; the application of fertilizer in crop cultivation is essential for profitable plant production. In particular, the intensive cultivation of vegetable crops involves high amounts of N fertilizer (as well as other nutrients) and often results in an excessive N supply [2]. This implies that vegetable production often does not meet the requirements of environmental legislations for good ecological water status (e.g., the EU Nitrates Directive or the EU Water Framework Directive). Vegetable production is, like no other, faced with the trade-off between ecological and economic demands. This adds to the importance for human health and global diet. Increasing the efficiency of nutrient supplies, especially nitrogen, is therefore essential for sustainable food production.

There is no general optimum for the nitrogen supply of vegetable crops; it depends on, among others, the production target (e.g., plant organ, quality). Consequently, the potential for a reduction in the N supply also varies among species and production systems. Generally, low N availability is known to decrease photosynthetic activity and thus plant growth [1]. Besides this, there are impacts on the secondary metabolism which can result

35

in decreased synthesis of N-based compounds such as alkaloids and increased synthesis of C-based secondary metabolites [3]. A limited N supply may consequently increase crops' quality, especially in terms of phytonutrients that are potentially beneficial for human health. In fact, a low N supply of plants is one possible way to achieve enrichment of secondary metabolites, as shown for several vegetable crops such as lettuce [4], and for medicinal plants such as chamomile [5]. The enrichment of secondary metabolites while maintaining high yield and quality may be particularly important for crops that are consumed in large quantities. Tomato offers great potential in this regard, as it is the most widely grown and consumed vegetable in the world [6]. A serving of tomatoes (defined as around 200 g of fresh tomatoes) covers the recommended dietary allowances of several mineral elements and, to a larger extent, those of vitamin C [7]. Due to the high consumption and concentration of many valuable compounds such as carotenes and polyphenols, tomato fruits are important sources of further secondary plant metabolites and can improve the intake of antioxidants in the human diet [8]. Carotenes such as lycopene and β-carotene are strong antioxidants and may reduce the risks for some cancers and coronary heart diseases, as well as those for diabetes and osteoporosis [9–11]. Polyphenols (such as flavonoids) are known to have antioxidant effects as well. Several studies suggest that polyphenols may protect against coronary heart diseases, counteract cancer development and prevent neurodegenerative diseases [12]. Additionally, tomato fruits are relevant sources of other antioxidants, such as ascorbic acid and vitamin E. An increase in the concentration of these secondary compounds in tomato fruits is therefore favorable for human health.

The internal quality and the taste of tomato fruits can be improved by agronomic strategies such as moderate salinity stress or modulated mineral nutrition [13]. When subjected to repeated short-term nitrogen limitation, tomato leaves contain increased flavonol glycoside concentrations [14]. Evidence for the transferability from tomato leaves to the fruits is still pending. Reduced N supply is known to cause an increase in soluble sugars and a decrease in acid concentrations of tomato fruits [15]. This may improve tomatoes' taste as sugars and acids are the major constituents of the taste experience [16]. Moreover, the N supply may alter other flavor components such as the volatile organic compounds of tomato fruits [17].

In tomato seedlings grown under N deficiency, shikimate and soluble sugars increased in the xylem sap, which suggests that the biosynthesis of phenolic substances might be favored [18]. In fact, increased concentrations of phenolic compounds in fruits resulted from reduced N supply at transplant or anthesis of greenhouse-grown tomatoes, albeit there were little effects on the carotenoid concentration [19].

However, the positive effects of reduced N supply on the secondary metabolism have not been found reliably. Bénard et al. [15] showed little effects of reduced N supply on secondary metabolites such as phenolic compounds, ascorbic acid and carotenoids in tomato fruits. Studies with two lycopene-rich tomato varieties revealed no impacts of reduced N supply on yield, firmness, valuable compounds and sensory properties of the fruits [20]. In contrast, Wang et al. [17] observed close relationships between N application rates and flavor parameters such as titratable acidity, soluble sugars, soluble solids, volatile compounds and fruit firmness in cherry tomatoes. Nevertheless, studies on quality and sensory properties of tomatoes grown with reduced N supply are rare, especially with the focus on shelf life.

This study aims at closing this gap in the knowledge. In a one-year trial, it was tested whether a reduction in the N supply by 50% of the recommended amount [21] impacts the yield and quality of tomato fruits. Fertilizers' reduction was performed at the beginning of fruit ripening to minimize the impact on fruit yield while maximizing the impact on fruit quality. Three varieties of cocktail tomato, differing in skin color, were chosen for the experiments since small-fruited and highly aromatic tomatoes are increasingly demanded by consumers. Quality parameters of the tomatoes were assessed after harvest and after eight-day storage.

We hypothesized that the reduced N supply (i) has no impact on the yield, firmness and color of tomato fruits, (ii) results in better taste due to the altered sugar-to-acid ratio of the fruits and (iii) enhances the nutritional quality of the fruits by increasing their antioxidant content.

2. Materials and Methods

2.1. Experimental Design and Plant Material

Three one-factorial experiments with the factor "N supply" and two factor levels, control (con) and reduced (Nred) N supplies, were performed for three cocktail tomato varieties separately in a greenhouse in Geisenheim, Germany. The experiments were set up in a randomized block design with four repetitions. Each repetition (plot) included ten plants. Subsequently, a storage experiment was carried out with the harvested fruits. Quality parameters were measured immediately after harvest and after storage. Hypotheses were tested with two-factorial analysis with the factors "N supply" and "storage".

For the greenhouse experiment, seeds of the tomato varieties 'Delioso' (red color; Rijk Zwaan Welver GmbH, Welver, Germany), 'Apresa' (orange color; Hild Samen GmbH, Marbach am Neckar, Germany) and 'Bombonera' (green-brownish color; Uniseeds Select B.V., Bleiswijk, The Netherlands) were sown into the substrate (Floradur A; BayWa AG, München, Germany) at the beginning of April 2012. The seedlings were picked and transferred into Ø 10 cm pots containing "Einheitserde ED Koko 1+1" (Einheitserdewerke Werkverband e.V., Sinntal-Altengronau, Germany) 14 d later. The transplants were watered and fertilized (Ferty 2 Mega, Planta Düngemittel GmbH, Regenstauf, Germany) as required. At the beginning of May 2012, the transplants (30 days old, BBCH stage 103) were planted into the soil (sandy loam) in the greenhouse with 2.8 plants m^{-2}. Plants of each tomato variety were planted onto eight randomly arranged plots (9.6 m^2 each with 10 plants per plot).

2.2. Cultivation and N Supply

The "con" treatment received weekly N supplies of 3 g m^{-2}, according to the recommendations for fertigated tomatoes [21]. For the "Nred" treatment, the weekly N supply was reduced by 50% at the onset of fruit ripening, yielding 1.5 g N m^{-2} per week.

Until the differentiated N application, the fertilizer "Ferty 2 Mega" was applied (1 g N m^{-2} per week; Hauert HBG Dünger AG, Grossaffoltern, Switzerland). Starting mid-May 2012, the N dose was increased to 2 g m^{-2} per week until early June 2012. At the breaker stage, when the fruits started to change color (early June 2012), the N control plots received 3 g N m^{-2} per week (composed of Ferty Basis 1, calcium nitrate and ammonium sulphate; Planta Düngemittel GmbH). The reduced N treatments received 50% less N, yielding 1.5 g N m^{-2} per week (composed of Ferty Basis 1 and calcium nitrate; Beiselen GmbH, Ulm, Germany). In total, plants received 360 (con) and 210 (Nred) kg N ha^{-1} throughout the growing season. Albeit this represents a reduction in total N application of approximately 42%, this is further regarded as a 50% reduction.

The fertilizer was applied with the irrigation water as fertigation. Each plant was supplied by drip irrigation (two drippers per plant, with 0.85 L h^{-1}), if soil tension fell below −9 kPa (measured at 20 cm soil depth by tensiometers). The application of water and fertilizer was controlled by the KLIWADU® software (developed by Michael Beck, Hochschule Weihenstephan-Triesdorf, Germany).

Pollination of the flowers was ensured by regular placement of bumblebees (re-natur GmbH, Stolpe, Germany) in the greenhouse. Plant protection was carried out by application of beneficial insects (*Amblyseius cucumeris* and *Encarsia formosa*, all from re-natur GmbH) when required.

The radiation inside the glass-covered greenhouse reached 724.7 kW m^{-2} during the cultivation period. The temperature regime throughout the cultivation season in the greenhouse is listed in Table 1.

Table 1. Temperature regime throughout the cultivation season.

Cultivation Stage	Minimum Day Temperature (°C)	Minimum Night Temperature (°C)	Temperature Set Point for Ventilation (°C)
Sowing till germination	24	20	26
Germination till planting	22	18	26
First week after planting	24	20	26
From second week after planting	20	16	22

2.3. Harvest

Mature tomato fruits were harvested at the fruit ripening stage "eating ripeness", which is chemometrically described as "red ripe" for red cultivars [22]. The different colored varieties were harvested analog to this. The harvest was carried out twice a week lasting from the end of June 2012 (week 26, corresponding to BBCH stage 801) to early September 2012 (week 36). The tomato fruits from each experimental plot were sorted into marketable and non-marketable fractions (broken, blossom end rot, green shoulder, soft, fissured, unfertilized, others). Fresh matter of tomato fruits was determined per plant and plot.

2.4. Storage

At least 12 marketable, similarly sized fruits from each experimental treatment were kept in a cold storage at 12 °C in the dark (CS 0300 T, Viessmann Kühlsysteme GmbH, Hof/Saale, Germany). The fruits were stored in plastic boxes in single layers and covered with plastic film to prevent water loss. After eight days, the fruits were removed from the cold storage for further analysis and sensory evaluations. During the course of the harvesting season, the storage experiment was repeated five times.

2.5. Spectroscopic Measurements

Non-invasive spectroscopic measurements were performed with the Multiplex® 3.6 device (Force-A, Orsay Cedex, France). Three marketable fruits from each plot were randomly chosen. Measurements were conducted in configuration 4 using red, blue, green and UV excitation at the bottom of the fruit. Parameters of interest were SFR_G (indicating the chlorophyll content), FLAV (indicating the concentration of flavonoids) and ANTH_RG (indicating the anthocyanin content) [23]. Averages of the indices of the fruits per plot were calculated for each date separately.

2.6. Firmness of the Fruit Pulp

For assessing the firmness of the fruit pulp, a single measurement was taken with the HHP-2001 device (Heinrich Bareiss Prüfgerätebau GmbH, Oberdischingen, Germany). A 0.25 cm^2 test anvil was pressed against the bottom of the fruit and after one second, resistance readings were taken. Average values per plot and measuring date were calculated.

2.7. Sampling and Chemical Analyses

After the spectroscopic measurements and the determination of fruit pulp firmness, the fruits were weighed, cut and dried at 60 °C. After three days, the dry matter was determined. The dried tomato fruits were sampled for analysis of minerals. For other chemical analyses, an aliquot of approximately 1000 g of fruits per plot was sampled and stored at −20 °C until further analyses.

Dried tomato samples were ground to a fine powder. For determination of mineral concentrations, an aliquot of the sample was extracted according to the Kjeldahl procedure (TURBOTHERM, C. Gerhardt GmbH & Co. KG, Königswinter, Germany). The extracts were filtered, and the content of N, P K, Ca, Mg, Fe, Zn, Mn and Cu was determined by inductively coupled plasma optical emission spectrometry (ICP-OES; SPECTRO ARCOS, SPECTRO Analytical Instruments, Kleve, Germany).

For further analyses, the frozen fruit material was thawed at room temperature for 12 h and then homogenized with commercial kitchen blenders. Aliquots of this homogenate were taken for carotenoid analysis. The remaining homogenate was centrifuged at 4500 rpm for 10 min (ROTANTA 460 RS, Andreas Hettich GmbH & Co. KG, Tuttlingen, Germany). The supernatant was filtered, and the resulting juice was kept deep-frozen at -20 °C until analysis of its composition.

The total content of liposoluble pigments (mainly carotenoids) from the tomato homogenate was extracted in acetone/hexane with addition of sodium sulphate and calcium carbonate (both of Sigma-Aldrich Chemie GmbH, Taufkirchen, Germany). The extracts were filtered (syringe filter ROTILABO PTFE 0.2 μm, Carl Roth GmbH + Co. KG, Karlsruhe, Germany) and the concentration of the liposoluble pigments was determined spectrophotometrically at 450 nm as equivalents of β-carotene. Calibration curves were conducted using β-carotene (Merck KGAa, Darmstadt, Germany) as a standard.

The juice was thawed for 3 h at room temperature prior to analysis of its composition. The following parameters were assessed according to the standards of the International Fruit Juice Union (Paris): soluble solids (refractometric analysis), conductivity (conductometric analysis), pH (potentiometric analysis), total acidity (expressed as citric acid, potentiometric analysis after titration to pH 8.1 with 0.3 M NaOH; Titroline alpha, Schott, Mainz, Germany), concentration of ascorbic acid (automatic potentiometric titration with SCHOTT TitroLine alpha plus, SI Analytics, Mainz, Germany). The contents of total phenolics were estimated by the Folin–Ciocalteu assay, and the antioxidative capacity was estimated by a photometric assay (as Trolox equivalent antioxidative capacity, TEAC). The color space parameters a* and b* were assessed by reflection readings of the juice against a white standard (CR 200, Konica Minolta Sensing Europe B.V., Bremen, Germany) and the "chroma" value was calculated according to D'Souza et al. [24].

All chemical analyses were performed as duplicates.

2.8. Sensory Evaluation

The effects of reduced N supply and the variety on the sensory attributes taste (sweetness, sourness, typically tomato, foreign taste, watery), consistency (firmness of fruit pulp, firmness of the peel, mealiness), aftertaste (intensity, foreign) and overall liking were assessed by quantitative descriptive sensory evaluation. The intensity of the attributes was rated on a scale from 1 (non-existent) to 9 (very strong). For assessing possible differences between the treatments before and after storage, extended triangle tests (as forced choice tests) were used. An additional question was posed concerning the preference of one sample. All samples were coded with three-digit random numbers by the "FIZZ Data Aquisition und Calculation" software (Biosystèmes, Couternon, France). For sensory evaluations, the tomato fruits were cut into four parts. Each sample consisted of six quarters from six different tomato fruits per treatment. The samples were presented in random order to the panelists. The sensory panel comprised 15 to 17 trained people that were asked to take bits from each tomato piece to obtain sensory impressions of mixed samples. The panelists received water and matzo (Dr. Schär AG/SPA, Burgstall, Italy) to clear their palates between the sensory evaluations of two samples.

2.9. Statistics

The impact of N supply, cold storage and their interaction was assessed by two-way-ANOVA for each tomato variety separately, if the conditions of normal distribution (tested with the Shapiro–Wilk test) and variance homogeneity (tested with the Levené test) were met. In case of no interactions between the factors N supply and storage, differences between the treatments were analyzed by Tukey pairwise comparison. If the data were not normally distributed or homogeneity of variances was not observed, the Kruskal–Wallis test followed by Mann–Whitney pairwise comparison was performed.

Impacts of the N supply on fruit yield data and on the content of mineral elements in the fruits were assessed by a *t*-test. All statistics of yield and quality data were performed (with $\alpha = 0.05$) using the PAST software (version 3.01, [25]).

The sensory data were analyzed by calculating averages for the single attributes and testing for significant differences with the Friedman test using the "FIZZ Data Aquisition und Calculation" software. When significant differences were observed, the means were separated by using least significant difference at $\alpha = 0.1$. The triangle tests were analyzed by comparing the percentage of correct answers to the probability of the corresponding binomial law.

3. Results

3.1. Yield

The total yield (=marketable and non-marketable fruits) was not significantly impacted by the N supply for the varieties 'Apresa' and 'Delioso' (Table 2). In contrast, 'Bombonera' had lower total fruit yields when supplied with reduced N amounts. Nevertheless, there were tendencies of a larger proportion of marketable fruits in the Nred treatment of 'Bombonera' ($p = 0.054$). For 'Delioso' and 'Bombonera', the fruits of the control treatments were significantly heavier compared to the Nred treatment (Table 2). This was not observed for the variety 'Apresa'.

Table 2. Total yield, proportion of marketable fruits and weight of single fruits in three tomato varieties with normal (con) and reduced (Nred) N supplies. $N = 4 \pm$ SD. Asterisks in the "Nred" line indicate significant differences between the "con" and the "Nred" treatment for the respective tomato varieties (*t*-test at $\alpha = 0.05$).

Variety	N Supply	Total Yield (kg m^{-2})	Proportion of Marketable Fruits (%)	Single Fruit Weight (g)
'Delioso'	con	4.27 ± 0.32	98.87 ± 0.51	39.67 ± 1.00
	Nred	3.98 ± 0.38	98.14 ± 0.86	36.75 ± 1.42 *
'Apresa'	con	3.13 ± 0.51	78.40 ± 0.71	33.88 ± 1.31
	Nred	2.91 ± 0.04	82.72 ± 2.68 *	32.77 ± 0.97
'Bombonera'	con	4.10 ± 0.07	36.59 ± 4.94	51.97 ± 1.18
	Nred	3.40 ± 0.23 *	50.31 ± 10.39	48.67 ± 1.71 *

3.2. External Quality Parameters

The firmness of the fruit pulp did not differ between the control and the Nred treatments of the three tomato varieties, both before and after storage (Table 3). However, tendencies of differences in the fruit pulp firmness due to different N supplies were observed in 'Bombonera' immediately after harvest ($p = 0.08$). In all varieties and N supply treatments, the fruit pulp firmness decreased significantly after storage.

The spectroscopic parameters FLAV, ANTH_RG and SFR_G are indices for the contents of different pigments of the fruit skin. Immediately after harvest, exclusively 'Bombonera' showed differences due to the N supply, with lower ANTH_RG indices in Nred compared to control fruits (Table 3). After storage, 'Apresa' fruits that were grown with reduced N supply had lower FLAV indices as compared to fruits grown under control conditions.

Fruits of the varieties 'Delioso' and 'Bombonera' did not show differences in the FLAV index between control and Nred treatments after storage. All indices were decreased after the storage in fruits of 'Delioso' grown under control and Nred conditions. The FLAV index was decreased after the storage in 'Apresa' control and Nred fruits but was not affected in 'Bombonera' fruits grown under both N supplies. In contrast, ANTH_RG was increased after storage in both control and Nred fruits of 'Bombonera' but was not affected in 'Apresa'. The index SFR_G was lowered after storage in 'Apresa' grown under reduced N supply, while there was no impact for 'Apresa' that was grown under control conditions.

Table 3. External quality parameters assessed by non-invasive measurements of tomato fruits of three cocktail varieties grown with normal (con) and reduced (Nred) N supplies, immediately after harvest (before storage) and after 8 days of storage at 12 °C (after storage). Means of N = 22 to 36 for firmness, and N = 31 to 58 $\pm$ SD for FLAV, ANTH_RG and SFR_G. Different letters indicate significant differences due to storage within a variety and N supply. Asterisks indicate significant differences due to N supply within a variety and storage (two-way ANOVA with Tukey pairwise comparison, or Kruskal–Wallis test with Mann–Whitney pairwise comparison; at α = 0.05).

Parameter (unit)	Storage	'Delioso' con	'Delioso' Nred	'Apresa' con	'Apresa' Nred	'Bombonera' con	'Bombonera' Nred
Firmness	Before	50.8 $\pm$ 4.3 a	50.7 $\pm$ 6.2 a	43.8 $\pm$ 4.8 a	43.9 $\pm$ 8.2 a	61.9 $\pm$ 6.0 a	58.3 $\pm$ 7.8 a
(kg cm^{-2})	After	46.1 $\pm$ 5.8 b	46.9 $\pm$ 4.2 b	40.9 $\pm$ 4.7 b	40.4 $\pm$ 5.4 b	53.7 $\pm$ 7.6 b	53.4 $\pm$ 8.0 b
FLAV	Before	1.48 $\pm$ 0.12 a	1.51 $\pm$ 0.15 a	1.62 $\pm$ 0.11 a	1.60 $\pm$ 0.10 a	2.19 $\pm$ 0.14 a	2.19 $\pm$ 0.06 a
	After	1.27 $\pm$ 0.16 b	1.26 $\pm$ 0.13 b	1.48 $\pm$ 0.12 b	1.44 $\pm$ 0.12 b *	2.19 $\pm$ 0.06 a	2.17 $\pm$ 0.07 a
ANTH_RG	Before	1.42 $\pm$ 0.07 a	1.40 $\pm$ 0.10 a	1.21 $\pm$ 0.12 a	1.20 $\pm$ 0.10 a	1.15 $\pm$ 0.16 a	1.22 $\pm$ 0.14 a *
	After	1.34 $\pm$ 0.14 b	1.33 $\pm$ 0.12 b	1.23 $\pm$ 0.07 a	1.20 $\pm$ 0.06 a	1.32 $\pm$ 0.13 b	1.34 $\pm$ 0.10 b
SFR_G	Before	1.26 $\pm$ 0.07 a	1.24 $\pm$ 0.07 a	1.35 $\pm$ 0.09 a	1.36 $\pm$ 0.11 a	1.28 $\pm$ 0.16 a	1.23 $\pm$ 0.12 a
	After	1.21 $\pm$ 0.08 b	1.20 $\pm$ 0.07 b	1.33 $\pm$ 0.08 a	1.31 $\pm$ 0.08 b	1.27 $\pm$ 0.14 a	1.26 $\pm$ 0.15 a

3.3. Chemical Composition of Tomato Fruits

The concentrations of the elements N, P, K, Mg, Fe, Zn, Mn and Cu differed between the varieties (data not shown) but were not influenced by N supply (Table 4).

Table 4. Mineral composition of tomato fruits at harvest. The concentrations of the mineral elements are expressed on a fresh weight base. Mean of N = 4 $\pm$ SD. No significant differences between the con and the Nred treatment for the respective tomato varieties were observed (t-test at α = 0.05).

Mineral	'Delioso' con	'Delioso' Nred	'Apresa' con	'Apresa' Nred	'Bombonera' con	'Bombonera' Nred
N (g kg^{-1})	2.90 $\pm$ 0.09	2.71 $\pm$ 0.23	3.15 $\pm$ 0.20	3.07 $\pm$ 0.35	2.87 $\pm$ 0.20	2.73 $\pm$ 0.07
P (g kg^{-1})	0.44 $\pm$ 0.02	0.44 $\pm$ 0.04	0.55 $\pm$ 0.02	0.54 $\pm$ 0.03	0.42 $\pm$ 0.04	0.40 $\pm$ 0.01
K (g kg^{-1})	5.05 $\pm$ 0.22	4.90 $\pm$ 0.41	6.36 $\pm$ 0.59	5.86 $\pm$ 0.46	5.29 $\pm$ 0.44	5.16 $\pm$ 0.28
Ca (g kg^{-1})	0.22 $\pm$ 0.03	0.21 $\pm$ 0.02	0.27 $\pm$ 0.06	0.21 $\pm$ 0.01	0.27 $\pm$ 0.09	0.20 $\pm$ 0.01
Mg (g kg^{-1})	0.19 $\pm$ 0.00	0.19 $\pm$ 0.02	0.24 $\pm$ 0.01	0.23 $\pm$ 0.01	0.22 $\pm$ 0.02	0.21 $\pm$ 0.01
Fe (mg kg^{-1})	5.88 $\pm$ 0.58	6.11 $\pm$ 0.67	7.78 $\pm$ 0.83	7.82 $\pm$ 1.75	5.91 $\pm$ 1.03	6.10 $\pm$ 0.73
Zn (mg kg^{-1})	3.60 $\pm$ 0.29	3.85 $\pm$ 0.53	4.40 $\pm$ 0.17	4.04 $\pm$ 0.40	5.34 $\pm$ 3.39	3.47 $\pm$ 0.09
Mn (mg kg^{-1})	1.12 $\pm$ 0.07	1.20 $\pm$ 0.09	1.41 $\pm$ 0.07	1.35 $\pm$ 0.10	1.29 $\pm$ 0.16	1.32 $\pm$ 0.09
Cu (mg kg^{-1})	1.79 $\pm$ 0.18	1.79 $\pm$ 0.26	2.16 $\pm$ 0.26	1.91 $\pm$ 0.26	1.46 $\pm$ 0.18	1.56 $\pm$ 0.08

The concentrations of total phenolics and total contents of liposoluble pigments were not affected by the N supply, both before and after storage, in all tomato varieties. Nevertheless, the eight-day storage of the fruits of 'Delioso' and 'Apresa' significantly increased their total content of liposoluble pigments (Table 5).

Ascorbic acid concentrations were higher in 'Delioso' fruits grown with reduced N supply both before and after storage. No changes in the ascorbic acid contents were observed in the fruits due to storage.

The antioxidant capacity (as TEAC) was higher in 'Delioso' fruits grown with reduced N supply before storage, but no differences between control and Nred fruits were observed after storage.

In general, the content of soluble solids, the pH and the total acidity of tomato fruits have impacts on their taste. The three parameters were not affected by the N supply before and after storage in fruits of all varieties here (Table 6). Storage had no effect on the content of soluble solids in all tested tomato samples but increased the pH in fruits of 'Delioso' and 'Apresa' grown under control and Nred conditions. There were no effects on the total acidity, except 'Delioso' under Nred conditions, where a decrease in total acidity was observed after storage. The fruits of 'Bombonera' showed no changes in the content of soluble solids, total acidity and pH due to N supply and storage.

Table 5. Mean and standard deviation of the antioxidant status of fruits of three cocktail tomato varieties grown with normal (con) and reduced (Nred) N supplies, immediately after harvest (before storage) and after 8 days of storage at 12 °C (after storage). The antioxidant parameters are expressed on a fresh weight base. N = 3–4. Different letters indicate significant differences due to storage within a variety and N supply. Asterisks indicate significant differences due to N supply within a variety and storage (Two-Way ANOVA with Tukey pairwise comparison, or Kruskal-Wallis test with Mann-Whitney pairwise comparison; at α = 0.05).

Parameter (unit)	Storage	'Delioso' con	'Delioso' Nred	'Apresa' con	'Apresa' Nred	'Bombonera' con	'Bombonera' Nred
Total phenolics (mg kg^{-1})	Before	355.5 ± 32.5 a	370.5 ± 35.3 a	410.3 ± 14.2 a	411.4 ± 17.7 a	343.3 ± 85.2 a	369.0 ± 54.1 a
	After	351.4 ± 24.8 a	382.5 ± 33.6 a	413.5 ± 20.6 a	420.5 ± 36.5 a	346.7 ± 5.4 a	355.9 ± 13.5 a
Total content of liposoluble pigments (mg kg^{-1})	Before	4.60 ± 0.20 a	4.45 ± 0.25 a	3.31 ± 0.24 a	3.00 ± 0.21 a	7.28 ± 0.15 a	7.04 ± 0.42 a
	After	5.75 ± 0.48 b	6.18 ± 0.26 b	4.19 ± 0.19 b	4.05 ± 0.16 b	7.05 ± 0.20 a	7.11 ± 0.13 a
Ascorbic acid (mg kg^{-1})	Before	170.3 ± 19.9 a	203.1 ± 12.0 a *	203.1 ± 64.1 a	191.2 ± 26.1 a	173.1 ± 26.6 a	167.8 ± 30.7 a
	After	185.8 ± 15.6 a	216.4 ± 15.9 a*	186.8 ± 17.8 a	188.1 ± 23.1 a	157.3 ± 2.8 a	147.9 ± 7.2 a
TEAC (mmol kg^{-1})	Before	3.10 ± 0.08 a	3.35 ± 0.10 a *	3.40 ± 0.27 a	3.55 ± 0.30 a	3.63 ± 0.32 a	3.67 ± 0.31 a
	After	3.28 ± 0.31 a	3.57 ± 0.19 b	3.79 ± 0.17 b	3.91 ± 0.28 a	3.36 ± 0.06 a	3.55 ± 0.32 a

Table 6. Composition and internal color (expressed as "chroma") of tomato fruits of three cocktail varieties grown with normal (con) and reduced (Nred) N supplies, immediately after harvest (before storage) and after 8 days of storage at 12 °C (after storage). Mean of N = 3–4 ± SD. Different letters indicate significant differences due to storage within a variety and N supply(two-way ANOVA with Tukey pairwise comparison, or Kruskal–Wallis test with Mann–Whitney pairwise comparison; at α = 0.05).

Parameter	Storage	'Delioso' con	'Delioso' Nred	'Apresa' con	'Apresa' Nred	'Bombonera' con	'Bombonera' Nred
Soluble solids (°Brix)	Before	5.40 ± 0.50 a	5.44 ± 0.40 a	5.73 ± 0.18 a	5.42 ± 0.46 a	5.53 ± 1.23 a	5.88 ± 0.76 a
	After	5.26 ± 0.37 a	5.38 ± 0.49 a	5.68 ± 0.23 a	5.48 ± 0.37 a	5.64 ± 0.07 a	5.74 ± 0.18 a
pH	Before	4.15 ± 0.02 a	4.16 ± 0.02 a	4.18 ± 0.02 a	4.16 ± 0.03 a	4.22 ± 0.05 a	4.18 ± 0.03 a
	After	4.24 ± 0.01 b	4.25 ± 0.03 b	4.25 ± 0.01 b	4.23 ± 0.03 b	4.22 ± 0.01 a	4.19 ± 0.04 a
Total acidity (g kg^{-1})	Before	4.13 ± 0.38 a	4.09 ± 0.21 a	4.96 ± 0.47 a	5.03 ± 0.22 a	4.08 ± 0.88 a	4.01 ± 0.54 a
	After	3.52 ± 0.27 a	3.57 ± 0.12 b	4.62 ± 0.27 a	4.31 ± 0.39 a	4.01 ± 0.03 a	3.95 ± 0.13 a
Chroma	Before	22.14 ± 0.33 a	21.85 ± 1.83 a	24.34 ± 1.31 a	25.37 ± 1.01 a	14.16 ± 4.97 a	10.28 ± 6.85 a
	After	19.10 ± 11.18 a	19.82 ± 11.92 a	4.83 ± 3.05 b	4.47 ± 2.93 b	10.33 ± 5.83 a	19.02 ± 1.58 a

The color index "chroma" was not significantly affected by both the N and storage treatments in all varieties of cocktail tomatoes (Table 6). Exclusively, 'Bombonera' grown with reduced N tended to have higher "chroma" values after storage compared to the control (p = 0.067). Due to storage, 'Apresa' grown under controlled and reduced N supplies showed a significant decrease in "chroma".

3.4. Sensory Evaluation

Descriptive sensory evaluation revealed differences in the attributes sweetness, watery, foreign aftertaste, firmness of the fruit pulp and the peel, typically tomato and overall liking in the three varieties (Figure 1a, statistical data not shown). Differences due to N supply were not observed for all attributes.

In triangle tests, the sensory panel was not able to discriminate significantly between controls and Nred samples immediately after harvest (Figure 1b). However, there was a trend for sensory discrimination between control and Nred treatments in 'Bombonera' fruits (p = 0.08). After storage, a significant difference between controls and Nred fruits was detected for 'Bombonera'. However, there was no preference for either controls or Nred fruits.

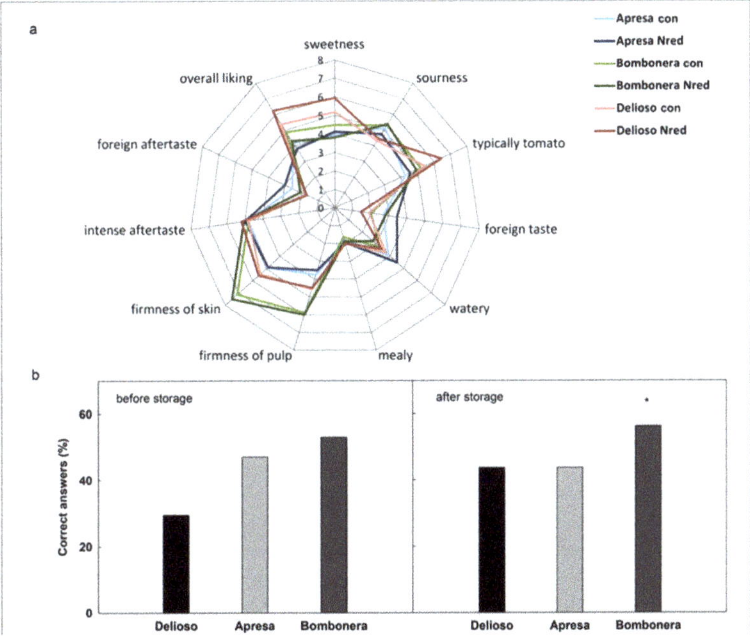

Figure 1. Sensory attributes of three tomato varieties grown with normal (con) and reduced (Nred) N supplies (**a**) and results of triangle tests (**b**). In (**b**), the "correct answers" indicate the relative abundance of correct distinctions between Nred and control samples. Asterisks in (**b**) show significant sensory distinctions between controls and Nred samples. $N = 16$ for (**a**). $N = 16$ to 17 for (**b**).

4. Discussion

Reducing the nitrogen supply in vegetable production aims at lowering the nitrate leaching into the groundwater and thus negative impacts to the environment. However, a sub-optimal nitrogen supply is known to alter the metabolism of plants and thereby their allocation of resources. As a consequence of this abiotic stress, photosynthetic and growth processes might be impeded, which could lower the crop yield [1]. In order to achieve environmentally sound fertilization while maintaining yield and quality, this experiment examined for tomato crops, cultivated in soil in a greenhouse, whether this balance could be achieved by reducing the N supply by 50% at the onset of fruit ripening. In fact, the total yield of fruits of the cocktail tomato varieties 'Delioso' and 'Apresa' (Table 2) was not reduced, being in accordance with studies using similar N levels for soil-grown tomato plants to our experiment (210 vs. 360 kg N ha^{-1}) [20,26–30].

The lower single fruit weight did not affect the total yield of 'Delioso', whereas this was not the case for 'Bombonera' (Table 2). Lower yields due to lower fruit weight with reduced N supply were reported by Qu et al. [31] and Wang et al. [32], but they occurred at lower N levels than in our experiment. It can be concluded that 'Bombonera' is more sensitive to reduced N availability under sub-optimal growing conditions as there were only around 37% marketable fruits in the control treatments (Table 2). Therefore, the reduction in the N supply was even beneficial for 'Bombonera' as the trend of increased fractions of marketable fruits resulted in comparable marketable yields of control and reduced N-grown fruits [33]. This positive effect of reduced N supply on the marketable fruit fraction was even significantly larger in 'Apresa' (Table 2) and was also observed in other studies, for example, due to lower numbers of rotten fruits [26,34].

Similarly to the marketable fruit yield, there was no impact of N reduction by 50% at the onset of fruit ripening on the firmness of the tomato fruits (Table 3). Albeit the fruit firmness decreased after storage in all varieties and treatments (as reported in other studies [35–38]), which—as consequence of decreased turgor and water loss by transpiration [39]—is a commonly known phenomenon of fruit ripening and senescence, there was no effect of the reduced N supply. This is surprising as N limitation is known to enhance the senescence processes of plants [1], but this is probably not true for fruit softening, as several other studies also report little effects on the firmness of tomato fruits [15,20,40–42].

Besides degradation of the cell wall polysaccharides, the ripening of tomato fruits is characterized by breakdown of chlorophylls in the peel [35] and concomitant increases in pigments such as carotenoids. These pigments possess antioxidative properties and scavenge reactive oxygen species which occur increasingly at the onset of fruit ripening [43], thereby accelerating senescence and thus being detrimental for the shelf life of tomatoes [44]. The concentrations of carbon-based carotenoids and flavonoids are known to increase in plant tissues with low N availability due to accumulation of carbohydrates that could not be used for growth processes [3]. This is supported by several studies that observed increasing concentrations in phenolic substances (i) in vegetative tissue [14,45–47], (ii) in fruits as a result of strongly reduced N supply (e.g., [19]), (iii) in soilless cultivation systems such as hydroponics (e.g., [7]) and (iv) at earlier developmental stages (at transplant, see [19]).

However, N supplies of 210 and 360 kg ha^{-1} did not impact the concentration of soluble solids and acidity after harvest in our experiment (Table 6), which is in accordance with other studies [20,26,32,42,48]. The fact that sweetness and acidity are the major constituents of the tomato taste experience, along with volatile aroma compounds and texture parameters [17,49], explains the lack of significant differences in the strength of selected attributes assessed by the sensory panel (Figure 1a).

As expected from the similar soluble solid contents of fruits of the control and reduced N treatments, suggesting that excess carbohydrates did not occur in the Nred fruits, neither differences in the concentrations of total phenolics and total contents of liposoluble pigments in the whole tomato fruits (Table 5), in the inner color index "chroma" (Table 6), nor in the index FLAV were found between control and Nred tomato fruits. The lack of differences between the index SFR_G of control and Nred fruits (Table 3) further suggests that the reduced N supply did not enhance chlorophyll breakdown and thus ripening and senescence processes. This is in accordance with other reports on the coloration of tomato fruits showing no effects of reduced N supply [15,41,50,51]. Similar to our findings, no impact of reduced N fertilization of tomato plants was observed on the content of the most important carotenoids in tomato fruits, lycopene, β-carotene and lutein [7,13,15,20,50,52].

In accordance with the very low effects of reduced N supply on phenolic substances (assessed by [19,20]), there were no impacts on the antioxidative capacity of water-soluble compounds (assessed as TEAC) of the fruits (Table 5), except 'Delioso' fruits. They showed higher TEAC values probably due to the higher contents of ascorbic acid (Table 5). Other studies showed that reductions in the N supply up to one third of the optimal level often had no effect on the vitamin C content of the tomato fruits [2,7,15,20,29,32]. Moreover, other results suggest that the antioxidative capacity of tomato fruits is little affected by the N supply [20,50,51].

Overall, there were very small effects of the reduced N supply on tomato fruit yield and quality. Even the N content of the fruits (Table 4) was not affected by the N supply, in the same way as the other mineral elements (Table 4), which is in accordance with other studies with a comparable or even more severe N reduction than in our experiment [26,29,40,42,53–55]. As proposed by Stefanelli et al. [3], reduced N contents in plants would be expected if the N supply to plants was low. However, the N content of the tomato leaves was not affected by the reduced N supply (data not shown), similarly to other reports of the N content in plants [40,53,54,56]. In other studies, there were no impacts of low N fertilizer levels on photosynthetic activity [57,58] and very little effects on the height and the above-ground biomass [27,40,53,56] of soil-grown tomato plants. This suggests that the reduced N supply in our experiment (210 kg N ha^{-1}) was not a limiting factor for the tomato plants in soil cultivation.

In fact, Hartz and Bottoms [28] stated that a "seasonal N rate of 200 kg ha^{-1} appeared adequate to maximize fruit yield", which is supported by various studies of tomatoes [29,34,41,48,56,59]. Moreover, Zotarelli et al. [27] proposed that N application rates above 176 kg ha^{-1} are not beneficial for tomato fruit yields but increase the leaching of nitrate. According to the classification of Hartz and Bottoms [28], our control treatments (360 kg N ha^{-1}) received excessive amounts of N and the tomato plants in our Nred treatments still had adequate N supplies, which might explain the lacking or minimal effects on tomato yield and quality.

These conclusions apply to tomato plants grown in soil, while soilless cultivation of tomato plants seems to increase their sensitivity towards reduced N supply as several studies report decreased photosynthetic activity [2], plant biomass [14,15,46,60] and fruit mineral contents [7]. However, it is possible to overcome N deficiency and ensure high fruit yields of soilless-grown tomato plants by supplying N levels larger than 4.5 mmol L^{-1} [2,15,19].

The very little effects of the N supply reduction by 50% at the onset of ripening were even less apparent after the storage. Exclusively, the higher ascorbic acid concentrations in 'Delioso', the lower FLAV index of 'Apresa' and the different taste observations in 'Bombonera' after the storage period could be ascribed to low N supply. All other storage effects, such as reduced firmness, decreased acidity and concomitant increase in pH and increases in liposoluble pigments such as carotenoids, are common phenomena that occur during ripening and senescence of tomato fruits [38]. However, in contrast to other observations [38,55,61], no increases in total soluble solid contents were observed in the three cocktail tomato varieties of our study.

Nevertheless, there was a trend (before storage) or even the significant perception (after storage) of different sensory properties between fruits of 'Bombonera' grown with either optimal or reduced N supply (Figure 1b). This sensory discrimination between the N supply levels can be less attributed to the soluble solid content and the total acidity than to the volatile organic compounds. It is possible that the concentration and composition of these compounds in tomato fruits were altered during the storage [62] and thus resulted in significantly different sensory properties of control and Nred-grown fruits of 'Bombonera' (Figure 1b).

Fruit storage at 12 °C for eight days had no effects on the content of ascorbic acid of all tomato varieties and treatments (Table 5) but increased the antioxidative capacity, significantly or in trend, in fruits of 'Apresa' and 'Delioso' (Table 5). This suggests that storage at 12 °C for eight days contributed to maintaining or even enhancing the potential nutritional quality of the tomato fruits, regardless of the N supply during their cultivation.

5. Conclusions

Reducing the N supply by 50% upon the onset of fruit ripening had little or no effects on the yield and external quality of three cocktail tomato varieties grown in soil. Contrary to our expectation, the products of secondary metabolism were not elevated, presumably due to the only moderately perceived stress. Reduced N supply thus did not result in increased nutritional quality and improved taste at harvest. However, fruit quality showed no deterioration with storage due to the lower N supply. The good shelf life was partly accompanied by increases in antioxidants. Consequently, the amount of nitrogen fertilizer can be reduced for several tomato varieties in soil cultivation while maintaining the yield, quality, taste and shelf life of tomato fruits.

Author Contributions: Conceptualization, L.S. and J.Z.; methodology, L.S.; investigation, L.S.; resources, J.Z.; data curation, L.S.; writing—original draft preparation, L.S. and J.Z.; writing—review and editing, L.S. and J.Z.; visualization, L.S.; supervision, J.Z. All authors have read and agreed to the published version of the manuscript.

Funding: This research received no external funding.

Data Availability Statement: The data presented in this study are available on request from the corresponding author. The data are not publicly available due to privacy.

Acknowledgments: The authors thank Ralf Schweiggert, Ralph Lehnart, Anja Giehl, Anja Rheinberger and Petra Kürbel for help with the chemical analyses; and Mirjam Hey for help with the sensory evaluation. Seeds were kindly provided by Rijk Zwaan Welver GmbH, Hild Samen GmbH and Uniseeds Select B.V.

Conflicts of Interest: The authors declare no conflict of interest.

References

1. Mu, X.; Chen, Y. The physiological response of photosynthesis to nitrogen deficiency. *Plant Physiol. Biochem.* **2021**, *158*, 76–82. [CrossRef] [PubMed]
2. Truffault, V.; Marlene, R.; Brajeul, E.; Vercambre, G.; Gautier, H. To Stop Nitrogen Overdose in Soilless Tomato Crop: A Way to Promote Fruit Quality without Affecting Fruit Yield. *Agronomy* **2019**, *9*, 80. [CrossRef]
3. Stefanelli, D.; Goodwin, I.; Jones, R. Minimal nitrogen and water use in horticulture: Effects on quality and content of selected nutrients. *Food Res. Int.* **2010**, *43*, 1833–1843. [CrossRef]
4. Bumgarner, N.R.; Scheerens, J.C.; Mullen, R.W.; Bennett, M.A.; Ling, P.P.; Kleinhenz, M.D. Root-zone temperature and nitrogen affect the yield and secondary metabolite concentration of fall- and spring-grown, high-density leaf lettuce. *J. Sci. Food Agric.* **2012**, *92*, 116–124. [CrossRef] [PubMed]
5. Kováčik, J.; Bačkor, M. Changes of phenolic metabolism and oxidative status in nitrogen-deficient *Matricaria chamomilla* plants. *Plant Soil* **2007**, *297*, 255–265. [CrossRef]
6. Statista. Erntemenge von Gemüse und Melonen weltweit nach Art im Jahr. 2014. Available online: www.statista.com/statistik/daten/studie/29305/umfrage/weltweite-erzeugung-von-gemuese-nach-arten/ (accessed on 28 September 2020).
7. Erba, D.; Casiraghi, M.C.; Ribas-Agustí, A.; Cáceres, R.; Marfà, O.; Castellari, M. Nutritional value of tomatoes (*Solanum lycopersicum* L.) grown in greenhouse by different agronomic techniques. *J. Food Compos. Anal.* **2013**, *31*, 245–251. [CrossRef]
8. Toor, R.K.; Savage, G.P. Antioxidant activity in different fractions of tomatoes. *Food Res. Int.* **2005**, *38*, 487–494. [CrossRef]
9. Tanaka, T.; Shnimizu, M.; Moriwaki, H. Cancer chemoprevention by carotenoids. *Molecules* **2012**, *17*, 3202–3242. [CrossRef]
10. Imran, M.; Ghorat, F.; Ul-Haq, I.; Ur-Rehman, H.; Aslam, F.; Heydari, M.; Shariati, M.A.; Okuskhanova, E.; Yessimbekov, Z.; Thiruvengadam, M.; et al. Lycopene as a Natural Antioxidant Used to Prevent Human Health Disorders. *Antioxidants* **2020**, *9*, 706. [CrossRef]
11. Przybylska, S. Lycopene—A bioactive carotenoid offering multiple health benefits: A review. *Int. J. Food Sci. Technol.* **2019**, *55*, 11–32. [CrossRef]
12. Halliwell, B. Dietary polyphenols: Good, bad, or indifferent for your health? *Cardiovasc. Res.* **2007**, *73*, 341–347. [CrossRef]
13. Flores, P.; Hernández, V.; Hellín, P.; Fenoll, J.; Cava, J.; Mestre, T.; Martínez, V. Metabolite profile of the tomato dwarf cultivar Micro-Tom and comparative response to saline and nutritional stresses with regard to a commercial cultivar. *J. Sci. Food Agric.* **2016**, *96*, 1562–1570. [CrossRef] [PubMed]
14. Larbat, R.; Olsen, K.M.; Slimestad, R.; Løvdal, T.; Bénard, C.; Verheul, M.; Bourgaud, F.; Robin, C.; Lillo, C. Influence of repeated short-term nitrogen limitations on leaf phenolics metabolism in tomato. *Phytochemistry* **2012**, *77*, 119–128. [CrossRef] [PubMed]
15. Bénard, C.; Gautier, H.; Bourgaud, F.; Grasselly, D.; Navez, B.; Caris-Veyrat, C.; Weiss, M.; Génard, M. Effects of low nitrogen supply on tomato (*Solanum lycopersicum*) fruit yield and quality with special emphasis on sugars, acids, ascorbate, carotenoids, and phenolic compounds. *J. Agric. Food Chem.* **2009**, *57*, 4112–4123. [CrossRef] [PubMed]
16. Klein, D.; Gkisakis, V.; Krumbein, A.; Livieratos, I.; Köpke, U. Old and endangered tomato cultivars under organic greenhouse production: Effect of harvest time on flavour profile and consumer acceptance. *Int. J. Food Sci. Technol.* **2010**, *45*, 2250–2257. [CrossRef]
17. Wang, Y.-T.; Huang, S.-W.; Liu, R.-L.; Jin, J.-Y. Effects of nitrogen application on flavor compounds of cherry tomato fruits. *J. Plant Nutr. Soil Sci.* **2007**, *170*, 461–468. [CrossRef]
18. Sung, J.; Sonn, Y.; Lee, Y.; Kang, S.; Ha, S.; Krishnan, H.B.; Oh, T.-K. Compositional changes of selected amino acids, organic acids, and soluble sugars in the xylem sap of N, P, or K-deficient tomato plants. *J. Plant Nutr. Soil Sci.* **2015**, *178*, 792–797. [CrossRef]
19. Hernández, V.; Hellín, P.; Fenoll, J.; Flores, P. Impact of nitrogen supply limitation on tomato fruit composition. *Sci. Hortic.* **2020**, *264*, 109173. [CrossRef]
20. Schmidt, L.; Hey, M.; Kürbel, P.; Zinkernagel, J. Einfluss verringerter Stickstoffdüngung auf Ertrag und Produktqualität von zwei lycopinreichen Tomatensorten. *DGG-Proceedings* **2013**, *3*, 5. [CrossRef]
21. Rather, K. Nährstoffversorgung von Tomaten in Bodenkultur unter dem Aspekt der Flüssigdüngung mit Tropfbewässerung. In *Aktuelle Versuchsergebnisse und Informationen Aus Baden-Württemberg*; LVG Heidelberg: Heidelberg, Germany, 2007; pp. 36–43.
22. Skolik, P.; Morais, C.L.M.; Martin, F.L.; McAinsh, M.R. Determination of developmental and ripening stages of whole tomato fruit using portable infrared spectroscopy and Chemometrics. *BMC Plant Biol.* **2019**, *19*, 236. [CrossRef] [PubMed]
23. Tremblay, N.; Wang, Z.; Cerovic, Z.G. Sensing crop nitrogen status with fluorescence indicators. A review. *Agron. Sustain. Dev.* **2012**, *32*, 451–464. [CrossRef]

24. D'Souza, M.C.; Singha, S.; Ingle, M. Lycopene Concentration of Tomato Fruit can be Estimated from Chromaticity Values. *HortScience* **1992**, *27*, 465–466. [CrossRef]
25. Hammer, O.; Harper, D.A.T.; Ryan, P.D. PAST: Paleontological Statistics Software Package for Education and Data Analysis. *Palaeontol. Electron.* **2001**, *4*, 9.
26. Ronga, D.; Pentangelo, A.; Parisi, M. Optimizing N Fertilization to Improve Yield, Technological and Nutritional Quality of Tomato Grown in High Fertility Soil Conditions. *Plants* **2020**, *9*, 575. [CrossRef]
27. Zotarelli, L.; Scholberg, J.M.; Dukes, M.D.; Muñoz-Carpena, R.; Icerman, J. Tomato yield, biomass accumulation, root distribution and irrigation water use efficiency on a sandy soil, as affected by nitrogen rate and irrigation scheduling. *Agric. Water Manag.* **2009**, *96*, 23–34. [CrossRef]
28. Hartz, T.K.; Bottoms, T.G. Nitrogen Requirements of Drip-irrigated Processing Tomatoes. *HortScience* **2009**, *44*, 1988–1993. [CrossRef]
29. Duan, P.; Sun, Y.; Zhang, Y.; Fan, Q.; Yu, N.; Dang, X.; Zou, H. Assessing the Response of Tomato Yield, Fruit Composition, Nitrogen Absorption, and Soil Nitrogen Fractions to Different Fertilization Management Strategies in the Greenhouse. *HortScience* **2019**, *54*, 537–546. [CrossRef]
30. Djidonou, D.; Lopiano, K.; Zhao, X.; Simonne, E.H.; Erickson, J.E.; Koch, K.E. Estimating nitrogen nutritional crop requirements of grafted tomatoes under field conditions. *Sci. Hortic.* **2015**, *182*, 18–26. [CrossRef]
31. Qu, Z.; Qi, X.; Shi, R.; Zhao, Y.; Hu, Z.; Chen, Q.; Li, C. Reduced N Fertilizer Application with Optimal Blend of Controlled-Release Urea and Urea Improves Tomato Yield and Quality in Greenhouse Production System. *J. Soil Sci. Plant Nutr.* **2020**, *20*, 1741–1750. [CrossRef]
32. Wang, C.; Gu, F.; Chen, J.; Yang, H.; Jiang, J.; Du, T.; Zhang, J. Assessing the response of yield and comprehensive fruit quality of tomato grown in greenhouse to deficit irrigation and nitrogen application strategies. *Agric. Water Manag.* **2015**, *161*, 9–19. [CrossRef]
33. Schmidt, L.; Zinkernagel, J. Is it possible to enhance the concentrations of valuable compounds by reduced nitrogen supply in several vegetable species? *Acta Hort.* **2014**, *1106*, 55–60. [CrossRef]
34. Dadomo, M.; Macua, J.I.; Gainza, A.M.; Christou, M.; Dumas, Y.; Branthôme, X.; Bussières, P. Influence of water and nitrogen availability on yield components of processing tomato in the European Union countries. *Acta Hort.* **1994**, *376*, 271–274. [CrossRef]
35. Paul, V.; Pandey, R. Role of internal atmosphere on fruit ripening and storability-A review. *J. Food Sci. Technol.* **2014**, *51*, 1223–1250. [CrossRef]
36. Verheul, M.J.; Slimestad, R.; Tjøstheim, I.H. From producer to consumer: Greenhouse tomato quality as affected by variety, maturity stage at harvest, transport conditions, and supermarket storage. *J. Agric. Food Chem.* **2015**, *63*, 5026–5034. [CrossRef]
37. Xu, S.; Sun, X.; Lu, H.; Yang, H.; Ruan, Q.; Huang, H.; Chen, M. Detecting and Monitoring the Flavor of Tomato (*Solanum lycopersicum*) under the Impact of Postharvest Handlings by Physicochemical Parameters and Electronic Nose. *Sensors* **2018**, *18*, 1847. [CrossRef]
38. Siddiqui, M.W.; Lara, I.; Ilahy, R.; Tlili, I.; Ali, A.; Homa, F.; Prasad, K.; Deshi, V.; Lenucci, M.S.; Hdider, C. Dynamic Changes in Health-Promoting Properties and Eating Quality During Off-Vine Ripening of Tomatoes. *Compr. Rev. Food Sci. Food Saf.* **2018**, *17*, 1540–1560. [CrossRef]
39. Lara, I.; Belge, B.; Goulao, L.F. The fruit cuticle as a modulator of postharvest quality. *Postharvest Biol. Technol.* **2014**, *87*, 103–112. [CrossRef]
40. Assunção, N.S.; Silva, N.O.; Fernandes, F.L.; de Aquino, L.A.; de Sena Fernandes, M.E. Physico-chemical characteristics and productivity of tomato plants in function of nitrogen sources and doses. *Biosci. J.* **2020**, *36*. [CrossRef]
41. Warner, J.; Zhang, T.Q.; Hao, X. Effects of nitrogen fertilization on fruit yield and quality of processing tomatoes. *Can. J. Plant Sci.* **2004**, *84*, 865–871. [CrossRef]
42. Maboko, M.M.; Du Plooy, C.P. Response of Hydroponically Grown Cherry and Fresh Market Tomatoes to Reduced Nutrient Concentration and Foliar Fertilizer Application under Shadenet Conditions. *HortScience* **2017**, *52*, 572–578. [CrossRef]
43. Corpas, F.J.; Freschi, L.; Rodríguez-Ruiz, M.; Mioto, P.T.; González-Gordo, S.; Palma, J.M. Nitro-oxidative metabolism during fruit ripening. *J. Exp. Bot.* **2018**, *69*, 3449–3463. [CrossRef] [PubMed]
44. Decros, G.; Baldet, P.; Beauvoit, B.; Stevens, R.; Flandin, A.; Colombié, S.; Gibon, Y.; Pétriacq, P. Get the Balance Right: ROS Homeostasis and Redox Signalling in Fruit. *Front. Plant Sci.* **2019**, *10*, 1091. [CrossRef]
45. Groher, T.; Schmittgen, S.; Noga, G.; Hunsche, M. Limitation of mineral supply as tool for the induction of secondary metabolites accumulation in tomato leaves. *Plant Physiol. Biochem.* **2018**, *130*, 105–111. [CrossRef] [PubMed]
46. Larbat, R.; Adamowicz, S.; Robin, C.; Han, P.; Desneux, N.; Le Bot, J. Interrelated responses of tomato plants and the leaf miner *Tuta absoluta* to nitrogen supply. *Plant Biol. (Stuttg)* **2016**, *18*, 495–504. [CrossRef]
47. Stewart, A.J.; Chapman, W.; Jenkins, G.I.; Graham, I.; Martin, T.; Crozier, A. The effect of nitrogen and phosphorus deficiency on flavonol accumulation in plant tissues. *Plant Cell Env.* **2001**, *24*, 1189–1197. [CrossRef]
48. Di Cesare, L.F.; Migliori, C.; Viscardi, D.; Parisi, M. Quality of tomato fertilized with nitrogen and phosphorus. *Ital. J. Food Sci.* **2010**, *22*, 186–191.
49. Vogel, J.T.; Tieman, D.M.; Sims, C.A.; Odabasi, A.Z.; Clark, D.G.; Klee, H.J. Carotenoid content impacts flavor acceptability in tomato (*Solanum lycopersicum*). *J. Sci. Food Agric.* **2010**, *90*, 2233–2240. [CrossRef]

50. Kaniszewski, S.; Kosson, R.; Grzegorzewska, M.; Kowalski, A.; Badełek, E.; Szwejda-Grzybowska, J.; Tuccio, L.; Agati, G. Yield and Quality Traits of Field Grown Tomato as Affected by Cultivar and Nitrogen Application Rate. *J. Agric. Sci. Technol.* **2019**, *21*, 683–697.

51. Frías-Moreno, M.N.; Espino-Díaz, M.; Dávila-Aviña, J.; Gonzalez-Aguilar, G.A.; Ayala-Zavala, J.F.; Molina-Corral, F.J.; Parra-Quezada, R.A.; Orozco, G.I.O. Preharvest nitrogen application affects quality and antioxidant status of two tomato cultivars. *Bragantia* **2020**, *79*, 134–144. [CrossRef]

52. Kim, Y.X.; Kwon, M.C.; Lee, S.; Jung, E.S.; Lee, C.H.; Sung, J. Effects of Nutrient and Water Supply During Fruit Development on Metabolite Composition in Tomato Fruits (*Solanum lycopersicum* L.) Grown in Magnesium Excess Soils. *Front. Plant Sci.* **2020**, *11*, 562399. [CrossRef]

53. Zotarelli, L.; Dukes, M.D.; Scholberg, J.; Muñoz-Carpena, R.; Icerman, J. Tomato nitrogen accumulation and fertilizer use efficiency on a sandy soil, as affected by nitrogen rate and irrigation scheduling. *Agric. Water Manag.* **2009**, *96*, 1247–1258. [CrossRef]

54. Geisseler, D.; Aegerter, B.J.; Miyao, E.M.; Turini, T.; Cahn, M.D. Nitrogen in soil and subsurface drip-irrigated processing tomato plants (*Solanum lycopersicum* L.) as affected by fertilization level. *Sci. Hortic.* **2020**, *261*, 108999. [CrossRef]

55. Tavallali, V.; Esmaili, S.; Karimi, S. Nitrogen and potassium requirements of tomato plants for the optimization of fruit quality and antioxidative capacity during storage. *Food Meas.* **2018**, *12*, 755–762. [CrossRef]

56. Sainju, U.M.; Dris, R.; Singh, B. Mineral nutrition of tomato. *Food Agric. Environ.* **2003**, *1*, 176–183.

57. Qiu, R.J.; Du, T.S.; Kang, S.Z.; Chen, R.Q.; Wu, L.S. Influence of Water and Nitrogen Stress on Stem Sap Flow of Tomato Grown in a Solar Greenhouse. *J. Amer. Soc. Hort. Sci.* **2015**, *140*, 111–119. [CrossRef]

58. Mazzitelli, J.B.; Lopez-Lauri, F.; Laurent, S.; Vidal, V.; Urban, L.; Gautier, H.; Gomez, L.; Vercambre, G.; Sérino, S.; Mazollier, C. Substantial Reduction of Water and Nitrogen Supply Does Not Negatively Affect Quality or Yield of Organically Grown Tomatoes. *Acta Hortic.* **2014**, *1041*, 103–108. [CrossRef]

59. Colla, G.; Battistelli, A.; Moscatello, S.; Proietti, S.; Casa, R.; Lo Cascio, B.; Leoni, C. Effects of reduced irrigation and nitrogen fertigation rate on yield, carbohydrate accumulation, and quality of processing tomatoes. *Acta Hortic.* **2001**, *542*, 187–196. [CrossRef]

60. Larbat, R.; Paris, C.; Le Bot, J.; Adamowicz, S. Phenolic characterization and variability in leaves, stems and roots of Micro-Tom and patio tomatoes, in response to nitrogen limitation. *Plant Sci.* **2014**, *224*, 62–73. [CrossRef]

61. Gómez, P.; Ferrer, M.Á.; Fernández-Trujillo, J.P.; Calderón, A.; Artés, F.; Egea-Cortines, M.; Weiss, J. Structural changes, chemical composition and antioxidant activity of cherry tomato fruits (cv. Micro-Tom) stored under optimal and chilling conditions. *J. Sci. Food Agric.* **2009**, *89*, 1543–1551. [CrossRef]

62. Krumbein, A.; Peters, P.; Brückner, B. Flavour compounds and a quantitative descriptive analysis of tomatoes (*Lycopersicon esculentum* Mill.) of different cultivars in short-term storage. *Postharvest Biol. Technol.* **2004**, *32*, 15–28. [CrossRef]

Article

For a Better Understanding of the Effect of N Form on Growth and Chemical Composition of C_3 Vascular Plants under Elevated CO_2—A Case Study with the Leafy Vegetable *Eruca sativa*

Lilian Schmidt [1,2] and Jana Zinkernagel [1,*]

1   Vegetable Crops, Hochschule Geisenheim University, Von-Lade-Strasse 1, 65366 Geisenheim, Germany;
     Lilian.Schmidt@hs-gm.de
2   Soil Science and Plant Nutrition, Hochschule Geisenheim University, Von-Lade-Strasse 1,
     65366 Geisenheim, Germany
*   Correspondence: Jana.Zinkernagel@hs-gm.de

Abstract: Plant responses to elevated atmospheric CO_2 (eCO_2) are well studied, but the interactions of the carbon and nitrogen metabolism in the process are still not fully revealed. This is especially true for the role of nitrogen forms and their assimilation by plants under eCO_2. This study investigated the interacting metabolic processes of atmospheric CO_2 levels and N form in the short-term crop arugula. The effects on physiological processes and their consequences for crop growth, yield and nutritional value were elucidated. Two varieties of arugula were grown in climate cabinets under 400 or 800 ppm CO_2, respectively. The plants were fertilized with either pure nitrate or ammonium-dominated-N. Photosynthetic CO_2 assimilation increased in response to eCO_2 regardless of the N form. This did not affect the assimilation of nitrate and consequently had no impact on the biomass of the plants. The extra photosynthates were not invested into the antioxidative compounds but were probably diverted towards the leaf structural compounds, thereby increasing dry mass and "diluting" several mineral elements. The fertilization of arugula with ammonium-dominated N had little benefits in terms of crop yield and nutritional quality. It is therefore not recommended to use ammonium-dominated N for arugula production under future elevated CO_2 levels.

Keywords: ammonium; climate change; food quality; photosynthesis; nitrogen source; nitrate; vegetable

Citation: Schmidt, L.; Zinkernagel, J. For a Better Understanding of the Effect of N Form on Growth and Chemical Composition of C_3 Vascular Plants under Elevated CO_2—A Case Study with the Leafy Vegetable *Eruca sativa*. *Horticulturae* **2021**, 7, 251. https://doi.org/10.3390/horticulturae7080251

Academic Editor: Pietro Santamaria

Received: 8 July 2021
Accepted: 13 August 2021
Published: 17 August 2021

Publisher's Note: MDPI stays neutral with regard to jurisdictional claims in published maps and institutional affiliations.

Copyright: © 2021 by the authors. Licensee MDPI, Basel, Switzerland. This article is an open access article distributed under the terms and conditions of the Creative Commons Attribution (CC BY) license (https://creativecommons.org/licenses/by/4.0/).

1. Introduction

The atmospheric CO_2 concentration is predicted to rise from 400 ppm up to 1142 ppm until the end of the 21st century [1]. Elevated CO_2 concentrations (eCO_2) promote growth for many plants as the water use efficiency is ameliorated and the carbon gain by photosynthetic activities is increased, at least in the short term, compared to ambient CO_2 concentrations (aCO_2) [2]). The beneficial effect of eCO_2 in higher plants, however, is linked to certain conditions, such as (e.g., [3]): i. the plant's type of photosynthesis (e.g., C_3 or C_4 pathway), ii. the plant's growth period and therefore effect duration (short-, medium- or long-term), iii. the presence of other photosynthesis limiting factors (e.g., light, nutrients), and iv. the balance between source activity (net export of photo-assimilates by mature leaves) and sink activity (growth, storage). This complexity highlights that the finely-tuned mechanism of carbon assimilation is the prerequisite for better plant production under future increased carbon dioxide supply. From an agricultural perspective, the physiological processes involved that can be influenced by crop management are of particular interest. The supply of nitrogen and its assimilation in plants represents one such adjusting screw. Nitrogen (N) is incorporated in many organic compounds comprising all amino acids, thus being highly relevant for the biosynthesis of enzymes which are essential for metabolic denwerke GmbH & Co. KG, Weilheim, Germany) with 30 seeds each. The substrate contained 41.6–49.8 mg total $N \cdot L^{-1}$. Sowing dates were 29 August 2016 for the first and second set of experiments and 24 October 2016 for the third set. The pots were covered with Styrofoam sheets and placed in trays into the climatic chambers set to the following conditions: 20/18 °C (day/night), no light, 400 ppm CO_2. After 4 days, the Styrofoam cover was removed. Another 3 days later, when the cotyledons were unfolded, the pots were arranged to their final position (distances of 11.5 cm × 14 cm). The photosynthetic photon flux density (metal halide lamps) was 1000 $\mu mol \cdot m^{-2} \cdot s^{-1}$ at the height of the plants during a photoperiod of 11 h (with dusk and dawn phases of one hour each). The light intensity at plant height was recorded by a quantum sensor (LI190R, LI-COR Environmental, Lincoln, NE, USA) coupled to a data logger (LI1400, LI-COR Environmental) in each climatic chamber separately. The relative humidity was set to 50% and the temperature to 20 °C/18 °C (day/night). Water was applied according to the demands of the single pots in order to fill up to a 90% water holding capacity per pot. The pots were thus weighed every 2 days at the beginning and daily later-on, and water was supplied when the water holding capacity fell below 80%.

Fertilizer was applied 4 times during the cultivation period, separately for N and P + K, yielding 300 mg N, 185 mg P_2O_5 and 500 mg K_2O per pot. N was given on days 6, 20, 27 and 33 after sowing as calcium nitrate (N + P + K = 15.5 + 0 + 0; 100% NO_3^-) in the pure nitrate treatment. On the same days, the ammonium-dominated N treatment received a mixture of 50% ammonium sulphate (N + P + K 21 + 0 + 0) and 50% ammonium nitrate (N + P + K 34,8 + 0 + 0), yielding 75% NH_4^+ and 25% NO_3^-. P and K was applied on days 13, 25, 29 and 35 after sowing ("Ferty Basisdünger 1" (N + P + K 0 + 14 + 38; Planta Düngemittel GmbH, Regenstauf, Germany). The fertilizer was added into the saucers in portions of 50–75 mL per pot. The plants were cultivated for 44 to 45 days. Harvests were conducted on 11 and 12 October 2016 (first and second set) and on 6 December 2016 (third set), respectively.

2.2. Biomass and Physiological Measurements

At harvest date, the number of inflorescences with at least 1 open flower and the total above-ground fresh mass were assessed from all 8 pots per treatment and climatic chamber (Figure 2). Three pots per treatment and climatic chamber were chosen for the recording of leaf numbers per pot, leaf area (LI 3100, LI-COR Biosciences, Lincoln, NE, USA), and the dry mass of the above-ground plant parts. Three other pots per treatment and climatic chamber were used for the measurement of photosynthetic gas exchange and non-invasive measurements of the N status and pigment indices. The above-ground biomass was dried at 60 °C for 24 h (when dry mass of the leaf material was constant) and then used for elemental analyses. The plants of 3 other pots per treatment and climatic chamber were used for the measurement of photosynthetic gas exchange. The above-ground biomass of all plants of these 3 pots per treatment and climatic chamber was sampled and frozen for later chemical analyses of plant pigments and vitamin C. The above-ground fresh mass in the remaining 2 pots per treatment and climatic chamber were sampled as well for the analysis of vitamin C and pigment content.

The gas exchange was measured with the GFS-3000 device (Heinz Walz GmbH, Effeltrich, Germany) on a single leaf per pot. The following cuvette conditions were used: flow 750 μmol, impeller 5, PAR 1100 $\mu mol \cdot m^{-2} \cdot s^{-1}$, relative humidity 50%, measured leaf area 8 cm^2, temperature 20 °C. The CO_2 supply was either set to 400 ppm or 800 ppm, depending on the respective growing conditions of the plants. Measuring points were recorded when the rates of transpiration and photosynthesis were stable. After 4 measurements, the gas analyzer was calibrated by taking measurements of an empty cuvette.

After the measurements, the leaves were cut at the edge of the cuvette and the edge was subtracted by cutting along a model made of a piece of paper. The remaining leaf part, which represented the actual measured leaf area, was photographed against a piece of paper of a defined size. The measured leaf area was calculated by colour segmentation with the software Fiji [45]. The data on photosynthetic gas exchange were then corrected for the actual measured leaf area.

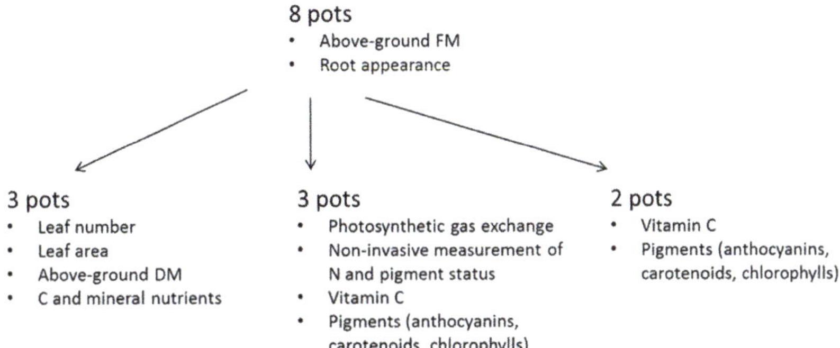

Figure 2. Parameters assessed at harvest of the arugula plants. Each pot contained 30 plants. FM = fresh mass, DM = dry mass.

On the same leaves as used for photosynthetic measurements, indices for the N status and pigments were assessed with the Dualex device (Force-A, Orsay Cedex, France) and the N-Tester (based on the SPAD device, YARA GmbH & Co. KG, Dülmen, Germany).

2.3. Chemical Analyses

For ascorbic acid analyses, the frozen leaf material was homogenized in oxalic acid and an aliquot was taken for quantification by titration with iodide and iodate (TitroLine alpha plus with automatic dispenser TITRONIC universal and sample changer TW alpha plus, all SCHOTT Instruments GmbH, Mainz, Germany). Chlorophylls and carotenoids were extracted from frozen and ground leaves in 100% acetone buffered with $NaHCO_3$ and quantified after photometric readings (modified procedure of [46]). Anthocyanins were analyzed according to [47]. Briefly, the anthocyanins were extracted from 0.5 g of frozen and ground leaf material using 80% methanol, which was acidified with 1% acetic acid. The extracts were measured photometrically at 530 nm and 657 nm and the concentrations of anthocyanins were calculated as equivalents to cyanidine-3-glycoside.

For elemental analysis, dried and ground plant materials were wet-digested in a microwave digester (MLS 1200 mega; MLS GmbH, Leutkirch, Germany). All micro- and macro-nutrients (excluding N & C) were analyzed using inductively coupled plasma optical emission spectrometry (Perkin-Elmer Optima 3000 ICP-OES; Perkin-Elmer Corp., Nerwalk, CT, USA). Total nitrogen was extracted using the Kjeldahl method and measured in a continuous flow apparatus (FIAstar™ 5000 Analyser; FOSS Analytical A/S, Hilleroed, Denmark). The organic C content in the dry material was measured by a colometric assay (Lambda 25 UV/VIS Spectrophotometer; PerkinElmer, Inc., Shelton, CT, USA) after wet oxidation of the organic matter using potassium dichromate [48].

The substrate of the 8 pots of the same treatment and the same climate chamber was combined and mixed before taking an aliquot as a single subsample. In this sample, the pH was recorded (inoLab pH 7310; Xylem Analytics Germany Sales GmbH & Co. KG, Weilheim, Germany). Additionally, the contents of the total N, NO_3^--N and NH_4^+-N were determined after Kjeldahl distillation (Vapodest; C. Gerhardt GmbH & Co. KG, Königswinter, Germany) and titration with 0.02 M HCl.

2.4. Statistical Analyses

The averages of the plants per climatic chamber and treatment were used for statistical analyses, yielding n = 3. The data assessed in a split plot design (CO_2 as the main factor with variety and N form as subfactors) were analyzed using the R software package [49]. The following model for a nested ANOVA was chosen: CO_2*N form*variety + error (climatic chamber/(N form*variety)). Following this, the interactions of the 3 experimental factors CO_2 concentration, N form and arugula variety were assessed. When significant impacts were observed, they were further dissected by *post hoc* F-tests at $\alpha = 0.05$.

3. Results

3.1. Substrate Composition

The total N remaining in the substrate after 44–45 days of arugula cultivation was neither impacted by CO_2 supply, nor by N form, nor by arugula variety (Table 1). The concentrations of ammonium-N and nitrate-N did not vary between the treatments. As well, the pH was not altered by the CO_2 supply (Table 1). However, the substrate of the 'Bellezia' plants had a slightly higher pH than those of the 'Tricia' plants. The N form also impacted the pH: with pure nitrate, the pH was higher when compared to ammonium-dominated N.

Table 1. Chemical parameters of the growth substrate at harvest (44–45 DAS). n = 3.

Treatment	Total N $(mg \cdot L^{-1})$	NH_4^+-N $(mg \cdot L^{-1})$	NO_3^--N $(mg \cdot L^{-1})$	pH
aBeNO	65.6 a	30.1 a	35.5 a	7.0 a
aTriNO	62.6 a	15.6 a	47.0 a	7.0 a
aBeNH	83.1 a	28.7 a	54.4 a	6.1 b
aTriNH	63.4 a	27.7 a	35.7 a	5.8 c
eBeNO	60.1 a	23.4 a	36.7 a	7.1 a
eTriNO	67.7 a	20.1 a	47.5 a	7.0 a
eBeNH	84.0 a	31.4 a	52.6 a	6.1 b
eTriNH	56.1 a	24.7 a	31.5 a	5.6 c

Different letters show differences between the experimental treatments at $\alpha = 0.05$. a = ambient CO_2, e = elevated CO_2, Be = 'Bellezia', Tri = 'Tricia', NO = pure nitrate, NH = ammonium-dominated N.

3.2. Plant Growth and Yield Parameters

The number of leaves per pot (=30 plants) was not significantly affected by the three experimental factors, albeit tendencies for more leaves in 'Bellezia' were obvious ($p = 0.1$). However, differences in the leaf area were detected, which consequently influenced the total leaf area per pot: The CO_2 concentration had no significant effect on the leaf area, albeit 'Bellezia' with ammonium-dominated N supply had a lower leaf area under elevated CO_2 as compared to ambient CO_2 concentrations. Thus, the interactions of N form, variety and CO_2 supply were apparent ($p = 0.048$). In general, plants with ammonium-dominated N supply, as well as those of the variety 'Bellezia', had lower leaf areas. The same pattern was found for the total above-ground fresh mass. While the CO_2 supply did not affect the fresh mass, there was an increase in the dry mass under elevated CO_2. The plants with pure nitrate had a higher dry mass. Concerning the arugula varieties, there were tendencies ($p = 0.053$) for larger above-ground dry mass in the 'Tricia' plants.

Inflorescences with at least one open flower showed up at day 39 after sowing (DAS), independently of CO_2 supply, N form and arugula variety (Figure S1). On 43 DAS, significant interactions of the three experimental factors' CO_2 supply, N form and arugula variety were observed ($p = 0.035$).

When fertilized with ammonium-dominated N, 'Bellezia' had less inflorescences than 'Tricia' under 400 ppm CO_2. This difference between the arugula varieties was not found at 800 ppm CO_2 when both 'Tricia' and 'Bellezia' with ammonium-dominated N supply had significantly more inflorescences than under ambient CO_2 concentrations. With pure nitrate, the number of inflorescences was not altered by the CO_2 supply in both arugula varieties. In general, no significant effects due to CO_2 were observed ($p > 0.1$, Table 2).

The N form exhibited significant effects on the number of inflorescences. At elevated CO_2 conditions, both arugula varieties had more inflorescences when grown with pure nitrate as compared to those with ammonium-dominated N.

Table 2. Above-ground biomass parameters at harvest. Data are averages of n = 3.

Treatment	Fresh Mass (g·plant^{-1})	Dry Mass (g·plant^{-1})	Leaf Number (plant^{-1})	Total Leaf Area (cm^2·plant^{-1})	Inflorescences (Number pot^{-1})
aBeNO	1.53 a	0.27 ac	8.11 a	25.48 ab	4.7 c
aTriNO	1.64 a	0.28 ad	7.81 a	30.10 c	4.1 c
aBeNH	1.11 b	0.18 b	8.09 a	21.25 de	2.6 d
aTriNH	1.28 ce	0.22 ab	7.70 a	25.84 ab	3.7 c
eBeNO	1.59 ad	0.32 cd	8.52 a	25.66 ab	6.5 ac
eTriNO	1.66 a	0.32 cd	8.20 a	28.30 ac	7.4 ac
eBeNH	1.14 bc	0.23 ab	8.46 a	19.04 d	5.8 b
eTriNH	1.30 e	0.26 a	8.19 a	24.03 be	5.9 b

Different letters indicate significant differences between the treatments at α = 0.05. a = ambient CO_2, e = elevated CO_2, Be = 'Bellezia', Tri = 'Tricia', NO = pure nitrate, NH = ammonium-dominated N.

3.3. Photosynthetic Gas Exchange

The photosynthetic CO_2 assimilation was not affected by N form and arugula variety but increased at eCO$_2$ (Figure 3a). However, eCO$_2$ significantly reduced the stomatal conductance (Figure 3b), thereby decreasing transpiration (data not shown), while increasing the leaf-internal CO_2 concentration (Figure 3c). The photosynthetic water use efficiency (WUE = assimilated CO_2 per water lost by transpiration) was 1.7 to 2.4 times higher at 800 ppm compared to 400 ppm CO_2 (Figure 3d).

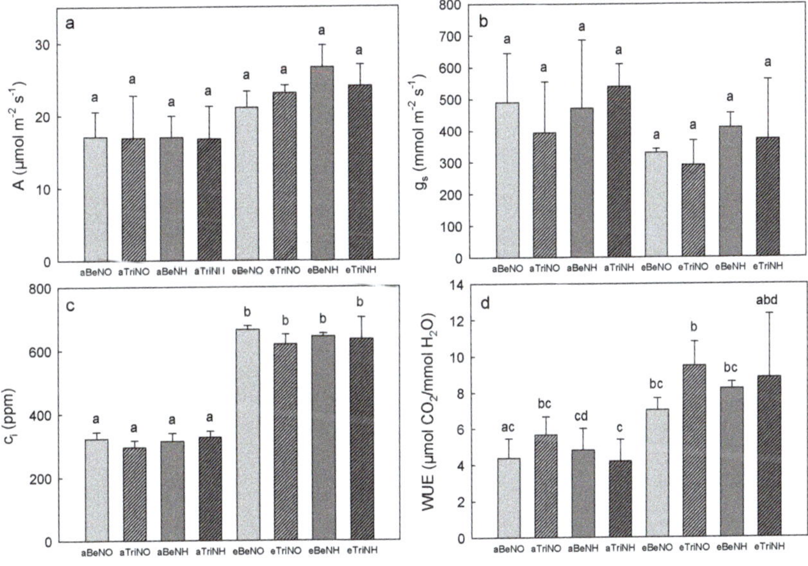

Figure 3. Photosynthetic CO_2 assimilation (**a**), stomatal conductance (**b**), leaf-internal CO_2 concentration (**c**) and photosynthetic water use efficiency (**d**) at harvest (45–46 days after sowing). a = ambient CO_2, e = elevated CO_2, Be = 'Bellezia', Tri = 'Tricia', NO = pure nitrate, NH = ammonium-dominated N. Data are averages of n = 3. Different letters indicate significant differences between the treatments at α = 0.05.

3.4. Plant Composition

The eight experimental treatments did not differ in terms of the chlorophyll content of the leaves, as confirmed by non-invasive measurements (Figure S2) and by chemical analyses (Table 3).

Table 3. Pigment and ascorbic acid contents of leaves of two arugula varieties grown with two different N forms and two atmospheric CO_2 concentrations. n = 3.

Treatment	Chlorophylls (mg 100 g^{-1} FM)	Carotenoids (mg 100 g^{-1} FM)	Anthocyanins (mg 100 g^{-1} FM)	Ascorbic Acid (mg 100 g^{-1} FM)
aBeNO	30.5 a	0.25 a	8.3 a	24.5 ab
aTriNO	29.8 a	0.16 a	8.1 a	18.1 a
aBeNH	37.5 a	0.18 a	9.4 a	18.1 a
aTriNH	29.1 a	0.15 a	6.8 a	21.5 ab
eBeNO	29.7 a	0.49 a	5.5 a	24.8 ab
eTriNO	25.5 a	0.40 a	6.2 a	21.8 ab
eBeNH	30.3 a	0.25 a	5.7 a	26.6 ab
eTriNH	30.1 a	0.32 a	6.4 a	35.2 b

Different letters indicate significant differences between the treatments at α = 0.05. a = ambient CO_2, e = elevated CO_2, Be = 'Bellezia', Tri = 'Tricia', NO = pure nitrate, NH = ammonium-dominated N.

The concentrations of total carotenoids, anthocyanins and ascorbic acid were not impacted by the experimental factors CO_2 concentration, N form and variety (Table 3). Ascorbic acid concentrations showed significant interactions of N form and variety (Table 3). The concentration of ascorbic acid of 'Tricia' increased by almost two-fold when grown with ammonium-dominated N at eCO_2 as compared to pure nitrate under aCO_2.

The N content of the leaves did not differ between the eight experimental treatments (Table 4). The elevated CO_2 concentrations tended to increase the C content of the arugula leaves (p = 0.09). Consequently the C/N ratio was significantly increased exclusively by 100% nitrate N, and in the variety 'Tricia' (data not shown).

Table 4. Concentration of macro- and micronutrients in above-ground dry mass of arugula. n = 3.

Element	Unit	aBeNO	aTriNO	aBeNH	aTriNH	eBeNO	eTriNO	eBeNH	eTriNH
C	%	39.1 a	37.8 a	38.1 a	38.6 a	39.2 a	39.5 a	39.9 a	39.6 a
N	%	2.19 a	2.16 a	2.97 a	2.47 a	1.92 a	1.85 a	2.61 a	2.51 a
S	%	0.96 ab	1.12 abc	1.31 b	1.20 ab	0.81 c	0.90 c	1.02 abc	0.94 ac
K	%	2.91 bc	3.18 bc	3.17 bc	3.30 b	2.68 ac	2.73 ac	2.94 bc	3.01 bc
P	%	0.39 a	0.36 a	0.71 b	0.61 bc	0.33 a	0.33 a	0.63 bc	0.57 c
Ca	%	1.87 a	1.97 a	1.65 a	1.37 a	1.73 a	1.74 a	1.38 a	1.33 a
Mg	%	0.22 a	0.24 a	0.22 a	0.22 a	0.21 a	0.22 a	0.21 a	0.21 a
Fe	ppm	41.4 a	42.7 a	60.1 a	71.5 a	39.4 a	39.1 a	71.9 a	63.6 a
Zn	ppm	20.3 a	21.0 abd	28.3 bc	30.9 c	20.5 ab	20.9 abd	31.1 c	28.3 cd
Mn	ppm	11.4 a	10.7 a	12.5 a	18.4 a	11.0 a	9.1 a	13.7 a	16.7 a
Cu	ppm	6.44 a	7.04 a	8.09 a	6.87 a	7.34 a	6.47 a	7.19 a	6.84 a
Na	ppm	0.16 ab	0.15 ab	0.22 a	0.15 ab	0.13 ab	0.13 b	0.15 ab	0.13 ab

Different letters indicate significant differences between the treatments at α = 0.05. a = ambient CO_2, e = elevated CO_2, Be = 'Bellezia', Tri = 'Tricia', NO = pure nitrate, NH = ammonium-dominated N.

The concentrations of the mineral elements Ca, Mg, Fe, Mn and Cu did not differ between the eight experimental treatments. The concentration of S was not altered by eCO_2, N form or arugula variety, with the exception of a decrease due to eCO_2 for 'Bellezia' fertilized with pure nitrate-N. Potassium concentrations were not altered by N form, variety and CO_2 concentration. However, 'Tricia' grown with ammonium-dominated N under aCO_2 had significantly larger K concentrations in the leaves as compared to 'Tricia' grown under eCO_2 with pure nitrate-N. The foliar concentrations of P were neither impacted by CO_2 concentration nor arugula variety. When supplied with ammonium-dominated N, the P concentrations increased in the leaves regardless of the CO_2 concentration. No effects of variety or CO_2 concentration on the Zn content of leaves were observed, except for increased contents in "eTriNH". For Na, no significant differences between the treatments were observed, albeit 'Bellezia' grown with ammonium-dominated N under aCO_2 (aBeNH) had larger Na concentrations than 'Tricia' grown under eCO_2 with pure nitrate-N (eTriNO; Table 4).

4. Discussion

This study is a contribution to the highly controversial debate of whether C_3 plants benefit from ammonium as a nitrogen source under elevated atmospheric CO_2 concentrations (eCO_2). This assumption is based on [50], who stated that nitrate assimilation in plants is reduced under non-CO_2-limited conditions, under which photorespiration is reduced in favour of carboxylation. Ammonium-N nutrition, compared to nitrate, would then be expected to increase the efficiency of carbon fixation and N assimilation, both leading to higher net production and product quality. However, Andrews et al. [17] argue against that theory with underestimated nitrate assimilation by roots under eCO_2 conditions. Besides the reduced leaf nitrate assimilation, the phenomenon of CO_2 acclimation, occurring potentially a few days or weeks after eCO_2 exposure [51], may mitigate the changes in the N metabolism and thus, the effect of the ammonium-nutrition of plants. Against this controversial background, we tested the effect of N form under eCO_2 on plant performance and leaf chemical composition using the short-term leafy vegetable arugula (*Eruca sativa*), which as a Brassicaceae highly contributes to human health.

When grown under 800 ppm CO_2, arugula plants showed a clear CO_2 fertilization effect as c_i (Figure 3c), photosynthetic CO_2 assimilation (Figure 3a) and dry mass (Table 2) increased in response to eCO_2, independently of the N form. These effects were recently confirmed for a close relative of arugula, *Arabidopsis thaliana* [52]. Characteristically for the CO_2 fertilization effect, the water use efficiency of arugula plants was improved (Figure 3d), as shown for several other C_3 species at eCO_2 [2].

The photosynthetic activity was enhanced, as there were no limitations by light, water supply, N supply or CO_2. The same is true for other mineral elements, such as Mg. As there were neither effects of the N form nor the CO_2 level on the Mg contents in the leaves (Table 4), we conclude that the carboxylation activity of Rubisco was favoured at the expense of the oxygenation of RuBP (see [4]).

This initially enhanced CO_2 assimilation was often shown to decrease under long-term eCO_2 exposure due to feedback inhibition by the accumulated carbohydrates in the leaves [53]. The so-called CO_2 acclimation effect was not observed for the short-term crop arugula (Figure 3a). Consequently, lower photorespiratory activities in plants under eCO_2 might be assumed compared to aCO_2, which is the prerequisite for limited nitrate assimilation in leaves [50].

However, our study cannot confirm Bloom's theory [50]. As nitrate assimilation was not quantified we used other methods, such as nitrate depletion from the growth medium which was considered to indicate altered nitrate assimilation in C_3 plants [4]. In fact, the lack of differences in soil nitrate-N and ammonium-N concentrations between all treatments at the end of the experiment suggests that nitrate depletion from the medium was unaffected by the CO_2 concentration (Table 1). Moreover, leaf N contents were similar in all N form*CO_2 treated plants (Table 4), which would not be expected with reduced leaf nitrate assimilation under eCO_2 levels. This further yielded similar fresh mass of the plants grown with ammonium-dominated N compared to pure nitrate-N under eCO_2 (Table 2). If nitrate assimilation in leaves was nevertheless reduced—and in fact there are indicators of reduced photorespiration and thus, limited nitrate-photoreduction—it must have been compensated by increased assimilation in other plant organs [17,54]. For *Arabidopsis thaliana*, it was suggested that root N assimilation is favoured in plants under eCO_2 conditions in order to "offset the decline in nitrogen metabolism in the leaves" [54]. A shift in the partitioning of nitrate assimilation consequently results in little alterations of the plant´s N status under eCO_2 conditions, which we thus can confirm for arugula.

For future fertilization under elevated CO_2, we do not recognize any advantage of ammonium-N nutrition for arugula in context of biomass and N accumulation. In general, using ammonium-dominated N is rather detrimental in arugula production as leaf area and fresh and dry mass may be decreased, regardless of the CO_2 level (Table 2). Moreover, smaller diameters and less visible roots [55] indicate overall restricted growth processes in response to ammonium-dominated N, pointing to mild toxicity symptoms [12].

This is especially true for the variety 'Bellezia' which probably is more sensitive towards ammonium-N supply than the variety 'Tricia'.

These negative aspects of ammonium-N nutrition can be counterbalanced by lower numbers of inflorescences at harvest (Table 2), especially under eCO_2 conditions, resulting in more marketable plants than with nitrate-N. This implies that senescence processes are not enhanced in response to the mild stress imposed by ammonium. Moreover, there was no increased exposure of plants to oxidative stress due to the lower photorespiratory activity under eCO_2 [4], as suggested by several antioxidants in the leaves (carotenoids, anthocyanins, ascorbic acid) that were not altered by CO_2 level and N form (Table 3). This is similar to results of other studies [50], but contrasts with reported increases in the contents of ascorbic acid and other antioxidants and their precursors in vegetables under eCO_2 [35,36,56–58].

Thus, the proposed carbon surplus generated by the increased photosynthetic activity under eCO_2 conditions was likely not utilized for increased synthesis of secondary metabolites with antioxidative properties [59]. In addition, C surplus was not increasingly metabolized to soluble carbohydrates in the leaves (as the C content was not significantly affected, Table 4), and thus did not cause feedback limitation of photosynthesis in the long-term (Figure 3a). The option of transporting extra carbon to the root system [59] can be excluded since the visible root growth was not promoted (see [55]). We therefore suggest that the surplus carbon was used for modifications of the leaves´ morphological and anatomical structures, as proposed by Gamage et al. [2]. This hypothesis is fostered by the significantly higher dry mass of the plants under eCO_2 levels (Table 2).

This higher dry mass resulted in the "dilution" of the macronutrients P, K and S (Table 4). The phenomenon of reduced concentrations of nitrogen and other mineral elements in plants grown under elevated CO_2 concentrations [35,39,40,57,60] is a consequence of accumulated photosynthates and reduced root uptake due to increased CO_2 assimilation and decreased transpiration, respectively [61]. In our study, the lower concentrations of relevant macronutrients in the above-ground biomass were apparently not limiting the plant growth processes, but they might lower the nutritional quality of the vegetables in the future [35,40].

This is especially true for the micronutrient Zn, as pointed out by Dong et al. [36] and Soares et al. [39]. However, foliar Zn levels were not affected by the CO_2 level in our study but were lower when grown with nitrate as the sole N source (Table 4). This can be regarded as a dilution effect by the larger biomass under pure nitrate fertilization.

To overcome these low contents of Zn in the leaves of arugula, the fertilization with ammonium-dominated N may be beneficial under ambient and elevated CO_2 (Table 4). However, this conclusion cannot be drawn for several other minerals (Ca, Mg, Na, Mn, Fe, Cu) for which no significant differences between the CO_2*N form treatments were observed (Table 4). The same applies to N (Table 4), which is closely linked to the leaf chlorophyll contents (Table 3). Several studies reported that the total chlorophyll content of leaves is not altered even at very high CO_2 levels [32,36,52,57,62], which enables the high photosynthetic capacities and the increased water use efficiency under non-limiting conditions, regardless of N form and arugula variety.

5. Conclusions

A CO_2 acclimation does not occur in the short-term leafy vegetable arugula. This crop benefits from elevated CO_2 levels in terms of photosynthetic activities and biomass development. The supply of ammonium-dominated N, in contrast to the hypothesis of Bloom [50], does not provide further advantages as the biomass development of arugula is rather impeded by this N form, regardless of the atmospheric CO_2 level. The same is true for the nutritional quality of arugula leaves, as pure nitrate supply and ammonium-dominated N fertilization have very similar impacts. It is thus not recommended to use ammonium-dominated N for the production of arugula under future elevated CO_2 levels.

Supplementary Materials: The following are available online at https://www.mdpi.com/article/10.3390/horticulturae7080251/s1, Figure S1: Number of inflorescences with at least one open flower observed from days 18 to 39 after the start of the experiments, Figure S2: Indices for the chlorophyll content of the leaves as observed by non-invasive measurements.

Author Contributions: Conceptualization, L.S. and J.Z.; Methodology, L.S. and J.Z.; Investigation, L.S.; Resources, J.Z.; Data Curation, L.S.; Writing—Original Draft Preparation, L.S. and J.Z.; Writing—Review & Editing, L.S. and J.Z.; Visualization, L.S.; Supervision, J.Z.; Project Administration, J.Z.; Funding Acquisition, J.Z. All authors have read and agreed to the published version of the manuscript.

Funding: This research was funded by the Hessian Ministry of Higher Education, Research, Science and the Arts in the framework "Forschung für die Praxis".

Institutional Review Board Statement: Not applicable.

Informed Consent Statement: Not applicable.

Data Availability Statement: The data presented in this study are available on request from the corresponding author. The data are not publicly available due to privacy.

Acknowledgments: The Alpenflor Erdenwerke GmbH & Co. KG donated the substrate, Enza Zaden donated the seeds. YARA provided the N-tester and the analytical data on the leaf mineral composition.

Conflicts of Interest: The authors declare no conflict of interest.

References

1. IPCC. Global Warming of 1.5 °C: An IPCC Special Report On the Impacts of Global Warming of 1.5 °C Above Pre-Industrial Levels and Related Global Greenhouse Gas Emission Pathways, in the Context of Strengthening the Global Response to the Threat of Climate Change, Sustainable Development, and Efforts to Eradicate Poverty. 2018. Available online: www.ipcc.ch/sr15/ (accessed on 7 July 2021).
2. Gamage, D.; Thompson, M.; Sutherland, M.; Hirotsu, N.; Makino, A.; Seneweera, S. New insights into the cellular mechanisms of plant growth at elevated atmospheric carbon dioxide concentrations. *Plant Cell Environ.* **2018**, *41*, 1233–1246. [CrossRef] [PubMed]
3. Ainsworth, E.A.; Rogers, A. The response of photosynthesis and stomatal conductance to rising CO_2: Mechanisms and environmental interactions. *Plant Cell Environ.* **2007**, *30*, 258–270. [CrossRef] [PubMed]
4. Bloom, A.J. Photorespiration and nitrate assimilation: A major intersection between plant carbon and nitrogen. *Photosynth. Res.* **2015**, *123*, 117–128. [CrossRef]
5. Stefanelli, D.; Goodwin, I.; Jones, R. Minimal nitrogen and water use in horticulture: Effects on quality and content of selected nutrients. *Food Res. Int.* **2010**, *43*, 1833–1843. [CrossRef]
6. Rubio-Asensio, J.S.; Bloom, A.J. Inorganic nitrogen form: A major player in wheat and *Arabidopsis* responses to elevated CO_2. *J. Exp. Bot.* **2017**, *68*, 2611–2625. [CrossRef] [PubMed]
7. Galmés, J.; Flexas, J.; Keys, A.J.; Cifre, J.; Mitchell, R.A.C.; Madgwick, P.J.; Haslam, R.P.; Medrano, H.; Parry, M.A.J. Rubisco specificity factor tends to be larger in plant species from drier habitats and in species with persistent leaves. *Plant Cell Environ.* **2005**, *28*, 571–579. [CrossRef]
8. Körner, C. Plant CO_2 responses: An issue of definition, time and resource supply. *New Phytol.* **2006**, *172*, 393–411. [CrossRef] [PubMed]
9. Bloom, A.J.; Asensio, J.S.R.; Randall, L.; Rachmilevitch, S.; Cousins, A.B.; Carlisle, E.A. CO_2 enrichment inhibits shoot nitrate assimilation in C_3 but not C_4 plants and slows growth under nitrate in C_3 plants. *Ecology* **2012**, *93*, 355–367. [CrossRef]
10. Bloom, A.J.; Burger, M.; Rubio Asensio, J.S.; Cousins, A.B. Carbon dioxide enrichment inhibits nitrate assimilation in wheat and *Arabidopsis*. *Science* **2010**, *328*, 899–903. [CrossRef]
11. Andrews, M.; Condron, L.M.; Kemp, P.D.; Topping, J.F.; Lindsey, K.; Hodge, S.; Raven, J.A. Elevated CO_2 effects on nitrogen assimilation and growth of C_3 vascular plants are similar regardless of N-form assimilated. *J. Exp. Bot.* **2019**, *70*, 683–690. [CrossRef]
12. Esteban, R.; Ariz, I.; Cruz, C.; Moran, J.F. Review: Mechanisms of ammonium toxicity and the quest for tolerance. *Plant Sci.* **2016**, *248*, 92–101. [CrossRef] [PubMed]
13. Britto, D.T.; Kronzucker, H.J. NH_4+ toxicity in higher plants: A critical review. *J. Plant Physiol.* **2002**, *159*, 567–584. [CrossRef]
14. Walch-Liu, P.; Neumann, G.; Bangerth, F.; Engels, C. Rapid effects of nitrogen form on leaf morphogenesis in tobacco. *J. Exp. Bot.* **2000**, *51*, 227–237. [CrossRef]
15. Hachiya, T.; Watanabe, C.K.; Fujimoto, M.; Ishikawa, T.; Takahara, K.; Kawai-Yamada, M.; Uchimiya, H.; Uesono, Y.; Terashima, I.; Noguchi, K. Nitrate addition alleviates ammonium toxicity without lessening ammonium accumulation, organic acid depletion and inorganic cation depletion in *Arabidopsis thaliana* shoots. *Plant Cell Physiol.* **2012**, *53*, 577–591. [CrossRef]

16. Torralbo, F.; González-Moro, M.B.; Baroja-Fernández, E.; Aranjuelo, I.; González-Murua, C. Differential regulation of stomatal conductance as a strategy to cope with ammonium fertilizer under ambient versus elevated CO_2. *Front. Plant Sci.* **2019**, *10*, 597. [CrossRef] [PubMed]
17. Andrews, M.; Condron, L.M.; Kemp, P.D.; Topping, J.F.; Lindsey, K.; Hodge, S.; Raven, J.A. Will rising atmospheric CO_2 concentration inhibit nitrate assimilation in shoots but enhance it in roots of C_3 plants? *Physiol. Plant* **2020**, *170*, 40–45. [CrossRef]
18. Rowland-Bamford, A.J.; Baker, J.T.; Allen, L.H.J.; Bowes, G. Acclimation of rice to changing atmospheric carbon dioxide concentration. *Plant Cell Environ.* **1991**, *14*, 577–583. [CrossRef]
19. Theobald, J.C.; Mitchell, R.A.C.; Parry, M.A.J.; Lawlor, D.W. Estimating the excess investment in ribulose-1,5-bisphosphate carboxylase/oxygenase in leaves of spring wheat grown under elevated CO_2. *Plant Physiol.* **1998**, *118*, 945–955. [CrossRef]
20. Moore, B.D.; Cheng, S.H.; Sims, D.; Seemann, J.R. The biochemical and molecular basis for photosynthetic acclimation to elevated atmospheric CO_2. *Plant Cell Environ.* **1999**, *22*, 567–582. [CrossRef]
21. Geiger, M.; Haake, V.; Ludewig, F.; Sonnewald, U.; Stitt, M. The nitrate and ammonium nitrate supply have a major influence on the response of photosynthesis, carbon metabolism, nitrogen metabolism and growth to elevated carbon dioxide in tobacco. *Plant Cell Environ.* **1999**, *22*, 1177–1199. [CrossRef]
22. Matt, P.; Geiger, M.; Walch-Liu, P.; Engels, C.; Krapp, A.; Stitt, M. Elevated carbon dioxide increases nitrate uptake and nitrate reductase activity when tobacco is growing on nitrate, but increases ammonium uptake and inhibits nitrate reductase activity when tobacco is growing on ammonium nitrate. *Plant Cell Environ.* **2001**, *24*, 1119–1137. [CrossRef]
23. Dong, J.; Li, X.; Chu, W.; Duan, Z. High nitrate supply promotes nitrate assimilation and alleviates photosynthetic acclimation of cucumber plants under elevated CO_2. *Sci. Hortic.* **2017**, *218*, 275–283. [CrossRef]
24. Bloom, A.J.; Kasemsap, P.; Rubio-Asensio, J.S. Rising atmospheric CO_2 concentration inhibits nitrate assimilation in shoots but enhances it in roots of C_3 plants. *Physiol. Plant* **2020**, *168*, 963–972. [CrossRef]
25. Kim, S.-J.; Chiami, K.; Ishii, G. Effect of ammonium: Nitrate nutrient ratio on nitrate and glucosinolate contents of hydroponically-grown rocket salad (*Eruca sativa* Mill.). *Soil Sci. Plant Nutr.* **2006**, *52*, 387–393. [CrossRef]
26. Albornoz, F. Crop responses to nitrogen overfertilization: A review. *Sci. Hortic.* **2016**, *205*, 79–83. [CrossRef]
27. Shang, H.Q.; Shen, G.M. Effect of ammonium/nitrate ratio on pak choi (*Brassica chinensis* L.) photosynthetic capacity and biomass accumulation under low light intensity and water deficit. *Photosynthetica* **2018**, *56*, 1039–1046. [CrossRef]
28. Santamaria, P.; Elia, A.; Papa, G.; Serio, F. Nitrate and ammonium nutrition in chicory and rocket salad plants. *J. Plant Nutr.* **1998**, *21*, 1779–1789. [CrossRef]
29. Santamaria, P.; Elia, A.; Parente, A.; Serio, F. Fertilization strategies for lowering nitrate content in leafy vegetables: Chicory and rocket salad cases. *J. Plant Nutr.* **1998**, *21*, 1791–1803. [CrossRef]
30. Zaghdoud, C.; Carvajal, M.; Moreno, D.A.; Ferchichi, A.; Del Carmen Martínez-Ballesta, M. Health-promoting compounds of broccoli (*Brassica oleracea* L. var. *italica*) plants as affected by nitrogen fertilisation in projected future climatic change environments. *J. Sci. Food Agric.* **2016**, *96*, 392–403. [CrossRef]
31. Marino, D.; Ariz, I.; Lasa, B.; Santamaria, E.; Fernandez-Irigoyen, J.; Gonzalez-Murua, C.; Aparicio Tejo, P.M. Quantitative proteomics reveals the importance of nitrogen source to control glucosinolate metabolism in *Arabidopsis thaliana* and *Brassica oleracea*. *J. Exp. Bot.* **2016**, *67*, 3313–3323. [CrossRef]
32. Chitarra, W.; Siciliano, I.; Ferrocino, I.; Gullino, M.L.; Garibaldi, A. Effect of elevated atmospheric CO_2 and temperature on the disease severity of rocket plants caused by fusarium wilt under phytotron conditions. *PLoS ONE* **2015**, *10*, e0140769. [CrossRef]
33. Gilardi, G.; Pugliese, M.; Chitarra, W.; Ramon, I.; Gullino, M.L.; Garibaldi, A. Effect of elevated atmospheric CO_2 and temperature increases on the severity of basil downy mildew caused by *Peronospora belbahrii* under phytotron conditions. *J. Phytopathol.* **2016**, *164*, 114–121. [CrossRef]
34. Eastburn, D.M.; McElrone, A.J.; Bilgin, D.D. Influence of atmospheric and climatic change on plant-pathogen interactions. *Plant Pathol.* **2011**, *60*, 54–69. [CrossRef]
35. Bisbis, M.B.; Gruda, N.; Blanke, M. Potential impacts of climate change on vegetable production and product quality—A review. *J. Clean. Prod.* **2018**, *170*, 1602–1620. [CrossRef]
36. Dong, J.; Gruda, N.; Lam, S.K.; Li, X.; Duan, Z. Effects of elevated CO_2 on nutritional quality of vegetables: A review. *Front. Plant Sci.* **2018**, *9*, 924. [CrossRef] [PubMed]
37. Muthusamy, M.; Hwang, J.E.; Kim, S.H.; Kim, J.A.; Jeong, M.-J.; Park, H.C.; Lee, S.I. Elevated carbon dioxide significantly improves ascorbic acid content, antioxidative properties and restricted biomass production in cruciferous vegetable seedlings. *Plant Biotechnol. Rep.* **2019**, *13*, 293–304. [CrossRef]
38. Jain, V.; Pal, M.; Raj, A.; Khetarpal, S. Photosynthesis and nutrient composition of spinach and fenugreek grown under elevated carbon dioxide concentration. *Biol. Plant* **2007**, *51*, 559–562. [CrossRef]
39. Soares, J.C.; Santos, C.S.; Carvalho, S.M.P.; Pintado, M.M.; Vasconcelos, M.W. Preserving the nutritional quality of crop plants under a changing climate: Importance and strategies. *Plant Soil* **2019**, *443*, 1–26. [CrossRef]
40. Loladze, I. Hidden shift of the ionome of plants exposed to elevated CO_2 depletes minerals at the base of human nutrition. *elife* **2014**, *3*, e02245. [CrossRef] [PubMed]
41. Hall, M.; Jobling, J.; Rogers, G. Some perspectives on rocket as a vegetable crop: A review. *Veg. Crop. Res. Bull.* **2012**, *76*, 334. [CrossRef]

42. Ramos-Bueno, R.P.; Rincón-Cervera, M.A.; González-Fernández, M.J.; Guil-Guerrero, J.L. Phytochemical composition and antitumor activities of new salad greens: Rucola (*Diplotaxis tenuifolia*) and corn salad (*Valerianella locusta*). *Plant Foods Hum. Nutr.* **2016**, *71*, 197–203. [CrossRef]
43. Herms, D.A.; Mattson, W.J. The dilemma of plants: To grow or defend. *Q. Rev. Biol.* **1992**, *67*, 283–335. [CrossRef]
44. Weinheimer, S.; Naab, B. Deutliche Unterschiede bei Rucola hinsichtlich der Anfälligkeit bei Falschem Mehltau und dem Schossverhalten. In *Versuche im Deutschen Gartenbau—Ökologischer Gemüsebau*; Verband der Landwirtschaftskammern e.V.: Berlin, Germany, 2014; pp. 230–232. (In German)
45. Schindelin, J.; Arganda-Carreras, I.; Frise, E.; Kaynig, V.; Longair, M.; Pietzsch, T.; Preibisch, S.; Rueden, C.; Saalfeld, S.; Schmid, B.; et al. Fiji: An open-source platform for biological-image analysis. *Nat. Methods* **2012**, *9*, 676–682. Available online: www.imagej.net/software/fiji/ (accessed on 7 July 2021). [CrossRef] [PubMed]
46. Ensminger, I.; Xyländer, M.; Hagen, C.; Braune, W. Strategies providing success in a variable habitat: III. Dynamic control of photosynthesis in *Cladophora glomerata*. *Plant Cell Environ.* **2001**, *24*, 769–779. [CrossRef]
47. Mancinelli, A.L. Interaction between light quality and light quantitiy in the photoregulation of anthocyanin production. *Plant Physiol.* **1990**, *92*, 1191–1195. [CrossRef] [PubMed]
48. Sims, J.R.; Haby, V.A. Simplified colorimetric determination of soil organic matter. *Soil Sci.* **1971**, *112*, 137–141. [CrossRef]
49. R Core Team. *R: A Language and Environment for Statistical Computing*; R Foundation for Statistical Computing: Vienna, Austria, 2015. Available online: www.R-project.org (accessed on 7 July 2021).
50. Bloom, A.J. Rising carbon dioxide concentrations and the future of crop production. *J. Sci. Food Agric.* **2006**, *86*, 1289–1291. [CrossRef]
51. Van Oosten, J.J.; Besford, R.T. Acclimation of photosynthesis to elevated CO_2 through feedback regulation of gene expression: Climate of opinion. *Photosynth. Res.* **1996**, *48*, 353–365. [CrossRef] [PubMed]
52. Dhami, N.; Cazzonelli, C.I. Short photoperiod attenuates CO_2 fertilization effect on shoot biomass in *Arabidopsis thaliana*. *Physiol. Mol. Biol. Plants* **2021**, *27*, 825–834. [CrossRef]
53. Tausz-Posch, S.; Tausz, M.; Bourgault, M. Elevated CO_2 effects on crops: Advances in understanding acclimation, nitrogen dynamics and interactions with drought and other organisms. *Plant Biol.* **2020**, *22* (Suppl. 1), 38–51. [CrossRef]
54. Jauregui, I.; Aparicio-Tejo, P.M.; Avila, C.; Cañas, R.; Sakalauskiene, S.; Aranjuelo, I. Root-shoot interactions explain the reduction of leaf mineral content in *Arabidopsis* plants grown under elevated CO_2 conditions. *Physiol. Plant.* **2016**, *158*, 65–79. [CrossRef] [PubMed]
55. Schmidt, L.; Zinkernagel, J. How to optimize the cultivation of arugula under elevated atmospheric CO_2 and different nitrogen forms? *DGG Proc.* **2018**, *8*, 1–5.
56. Jin, C.W.; Du, S.T.; Zhang, Y.S.; Tang, C.; Lin, X.Y. Atmospheric nitric oxide stimulates plant growth and improves the quality of spinach (*Spinacia oleracea*). *Ann. Appl. Biol.* **2009**, *155*, 113–120. [CrossRef]
57. Dong, A.; Armstrong, B.; Rajashekar, C.B. Elevated carbon dioxide level suppresses nutritional quality of lettuce and spinach. *Am. J. Plant Sci.* **2016**, *7*, 246–258. [CrossRef]
58. Becker, C.; Kläring, H.-P. CO_2 enrichment can produce high red leaf lettuce yield while increasing most flavonoid glycoside and some caffeic acid derivative concentrations. *Food Chem.* **2016**, *199*, 736–745. [CrossRef] [PubMed]
59. Prescott, C.E.; Grayston, S.J.; Helmisaari, H.-S.; Kaštovská, E.; Körner, C.; Lambers, H.; Meier, I.C.; Millard, P.; Ostonen, I. Surplus carbon drives allocation and plant-soil interactions. *Trends Ecol. Evol.* **2020**, *35*, 1110–1118. [CrossRef]
60. Loladze, I. Rising atmospheric CO_2 and human nutrition: Toward globally imbalanced plant stoichiometry? *Trends Ecol. Evol.* **2002**, *17*, 457–461. [CrossRef]
61. Taub, D.R.; Wang, X. Why are nitrogen concentrations in plant tissues lower under elevated CO_2? A critical examination of the hypotheses. *J. Integr. Plant Biol.* **2008**, *50*, 1365–1374. [CrossRef]
62. Sato, S.; Yanagisawa, S. Characterization of metabolic states of *Arabidopsis thaliana* under diverse carbon and nitrogen nutrient conditions via targeted metabolomic analysis. *Plant Cell Physiol.* **2014**, *55*, 306–319. [CrossRef]

MDPI

Article

Combined Effect of Salinity and LED Lights on the Yield and Quality of Purslane (*Portulaca oleracea* L.) Microgreens

Almudena Giménez [1], María del Carmen Martínez-Ballesta [1,2], Catalina Egea-Gilabert [1,2], Perla A. Gómez [1,2], Francisco Artés-Hernández [1,2], Giuseppina Pennisi [3], Francesco Orsini [3], Andrea Crepaldi [4] and Juan A. Fernández [1,2,*]

[1] Department of Agronomical Engineering, Universidad Politécnica de Cartagena, 30203 Cartagena, Spain; almudena.gimenez@upct.es (A.G.); mcarmen.ballesta@upct.es (M.d.C.M.-B.); catalina.egea@upct.es (C.E.-G.); perla.gomez@upct.es (P.A.G.); fr.artes-hdez@upct.es (F.A.-H.)
[2] Institute of Plant Biotechnology, Universidad Politécnica de Cartagena, Cartagena, 30202 Murcia, Spain
[3] Department of Agricultural and Food Sciences and Technologies, Alma Mater Studiorum, Università di Bologna, 40127 Bologna, Italy; giuseppina.pennisi@unibo.it (G.P.); f.orsini@unibo.it (F.O.)
[4] Flytech s.r.l., Via Dell'Artigianato 65, 32010 Paludi, Italy; andrea.crepaldi@flytech.it
* Correspondence: juan.fernandez@upct.es

Abstract: The present work aims to explore the potential to improve quality of purslane microgreens by combining water salinity and LED lighting during their cultivation. Purslane plants were grown in a growth chamber with light insulated compartments, under different lighting sources on a $16\,\text{h}\,\text{d}^{-1}$ photoperiod—fluorescent lamps (FL) and two LED treatments, including a red and blue (RB)) spectrum and a red, blue and far red (RB+IR) LED lights spectrum—while providing all of them a light intensity of $150\,\mu\text{mol}\,\text{m}^{-2}\,\text{s}^{-1}$. Plants were exposed to two salinity treatments, by adding 0 or 80 mM NaCl. Biomass, cation and anions, total phenolics (TPC) and flavonoids content (TFC), total antioxidant capacity (TAC), total chlorophylls (Chl) and carotenoids content (Car) and fatty acids were determined. The results showed that yield was increased by 21% both in RB and RB+FR lights compared to FL and in salinity compared to non-salinity conditions. The nitrate content was reduced by 81% and 91% when microgreens were grown under RB and RB+FR, respectively, as compared to FL light, and by 9.5% under saline conditions as compared with non-salinity conditions. The lowest oxalate contents were obtained with the combinations of RB or RB+FR lighting and salinity. The content of Cl and Na in the leaves were also reduced when microgreens were grown under RB and RB+FR lights under saline conditions. Microgreens grown under RB light reached the highest TPC, while salinity reduced TFC, Chl and Car. Finally, the fatty acid content was not affected by light or salinity, but these factors slightly influenced their composition. It is concluded that the use of RB and RB+FR lights in saline conditions is of potential use in purslane microgreens production, since it improves the yield and quality of the product, reducing the content of anti-nutritional compounds.

Keywords: minerals; fatty acids; oxalate; nitrate; phytochemicals; antioxidants

Citation: Giménez, A.; Martínez-Ballesta, M.d.C.; Egea-Gilabert, C.; Gómez, P.A.; Artés-Hernández, F.; Pennisi, G.; Orsini, F.; Crepaldi, A.; Fernández, J.A. Combined Effect of Salinity and LED Lights on the Yield and Quality of Purslane (*Portulaca oleracea* L.) Microgreens. *Horticulturae* **2021**, 7, 180. https://doi.org/10.3390/horticulturae7070180

Academic Editor: Lilian Schmidt

Received: 8 June 2021
Accepted: 25 June 2021
Published: 3 July 2021

Publisher's Note: MDPI stays neutral with regard to jurisdictional claims in published maps and institutional affiliations.

Copyright: © 2021 by the authors. Licensee MDPI, Basel, Switzerland. This article is an open access article distributed under the terms and conditions of the Creative Commons Attribution (CC BY) license (https://creativecommons.org/licenses/by/4.0/).

1. Introduction

Since antiquity, purslane (*Portulaca oleracea* L.) has been used as a medicinal and edible plant. As an edible plant, the most frequent use of purslane is raw in mixed salads, being appreciated for its succulence and pleasant, slightly acidic taste [1]. Its morphological and nutritional characteristic makes it suitable as a ready-to-eat product that can be easily adjusted to new market requirements [2]. Furthermore, it presents a higher nutritional value as compared with other major cultivated vegetables, thanks to the numerous bioprotective compounds such as antioxidants and vitamins, essential amino acids, omega-3 fatty acids and several minerals, including potassium [3]. On the other hand, its acceptance in the human diet could be conditioned by the large amounts of some anti-nutritional compounds it contains, mainly oxalates and nitrates, an aspect to be considered in the cultivar selection [2] and ameliorated by the appropriate choice of the growing season [4]. In recent years, together with the rise in its consumption, an increase in the agricultural area devoted to this crop has been also observed, particularly in the Mediterranean basin [5].

Purslane has been rated as moderately tolerant to salinity [6] by accumulating salt in the tissues [7]. In general, purslane plants exposed to salinity stress were shown to increase the concentration of nutritional compounds, including total phenol contents (TPC), total flavonoid contents (TFC) and ferric reducing antioxidant power [8]. Moreover, leaf nitrate concentration was decreased in purslane plants grown under salinity above 5.0 dS m^{-1} in the nutrient solution [9]. When purslane plants were treated with moderate salinity (up to 40 mM of NaCl in nutrient solution), no detrimental effects were observed in the biomass production, while an increase in the major fatty acids (linoleic, linolenic and palmitic acid) and a decrease in oxalic acid contents was measured [10]. Furthermore, plants exposed to four levels of salinity (60, 90, 120 and 240 mM NaCl) in the root zone for 40 days showed no signs of toxicity [11]. However, different accessions exhibit diverse performances under exposure to different levels of salinity stress, with evidence of ornamental purslane being generally more salt tolerant than common purslane [12].

Light-emitting diodes (LEDs) are increasingly adopted to efficiently provide lighting for production of several vegetable modalities and for quality preservation during storage, with proven potentialities to influence the metabolic pathways and therefore the biosynthesis of several bioactive compounds [13–15]. Fully controlled growing environments with artificial lighting (e.g., plant factories, vertical farms, growing chambers) are commonly used for microgreens cultivation. In these environments, light parameters (e.g., spectral quality, light intensity, photoperiod) can be optimized in order to enhance the nutritional value of microgreens [16,17]. Particularly, responses in terms of crop productivity and functional quality to changes in light spectrum depends on plant species [18], and for new and emerging species, including purslane, these effects are not yet well established. Recently, Kyriacou et al. [19] studied the effects of different spectral bandwidths on the nutritive and phytochemical composition of purslane, demonstrating that the use of monochromatic blue or red light resulted in higher accumulation of nitrate and lower biosynthesis of total polyphenols as compared to a combination of red and blue light. However, to our knowledge, no published information is available about the combined effects of salinity and LED spectral quality on the bioactive compounds' accumulation of purslane microgreens. Therefore, the aim of this study was to analyze the effects of salinity (80 mM NaCl) and LED spectral quality on yield and quality of purslane microgreens production. For that, biomass production, cations and anions content, fatty acids, and the main total bioactive compounds contents were determined in untreated and salt-treated plants grown under fluorescent light or two LED lighting treatments.

2. Materials and Methods

2.1. Plant Material and Growing Conditions

Seeds of common purslane (*Portulaca oleracea* L.), "Summer Purslane" (Tozer Seeds Ltd., Cobham, UK), were soaked in 0.5% sodium hypochlorite for 3 h under aeration and rinsed three times in distilled water. After sterilization, the seeds were evenly spread in 10 × 15 cm trays (0.3 g per tray) lined with autoclaved cellulose growth pads of white viscose (CN Seeds, Ely, UK), and immediately placed in three separated compartments for each light treatment in a plant growth chamber with alternating day/night temperatures of 25 °C/20 °C and a constant relative humidity (80%). Lighting treatments were kept throughout the experiment. A week later, half of the trays for each light treatment were irrigated with 20 mL per tray of a nutrient solution [20] once a week for 3 weeks, and the other half with a saline treatment, where 80 mM of NaCl were added to the nutrient solution. Microgreens were harvested and weighted when seedlings had four true leaves.

2.2. Lighting Treatments

The experiment included three lighting treatments, namely a control (FL) provided by fluorescent lamps (Philips 36W/54-765) and two LED treatments, providing a red and blue spectrum (RB) and a red, blue and far-red spectrum (RB+FR), respectively (Figure 1). LED lamps used for the study were manufactured by Flytech s.r.l. (Belluno, Italy) and featured diodes with narrow bands in the Hyper Red (R, peak at 669 nm), Blue (B, peak at 465 nm) and Far-Red (FR, peak at 730 nm) spectral regions. The RB treatment featured a red:blue ratio of 3, as formerly suggested in a wide range of vegetables by Pennisi et al. [13,21] The RB+FR treatment additionally integrated a 15% far red (FR) radiation. Lamps used for the FL treatment featured an RB ratio of 0.74. Spectral properties were determined using an illuminance spectrophotometer (CL-500A, Konica Minolta, Chiyoda, Tokyo, Japan). In all treatments, a constant photosynthetic photon flux density (PPFD) of 150 ± 5 μmol m^{-2} s^{-1} over the plant canopy was provided, and measured using a QSO (Apogee instruments, Logan, UT, USA) photosynthetic active radiation (PAR) Photon Flux Sensor (with equal sensitivity to red and blue radiation), connected with a ProCheck handheld reader (Decagon Devices Inc., Pullman, WA, USA) [13]. A photoperiod of 16 h light/8 h darkness was used in all the lighting treatments.

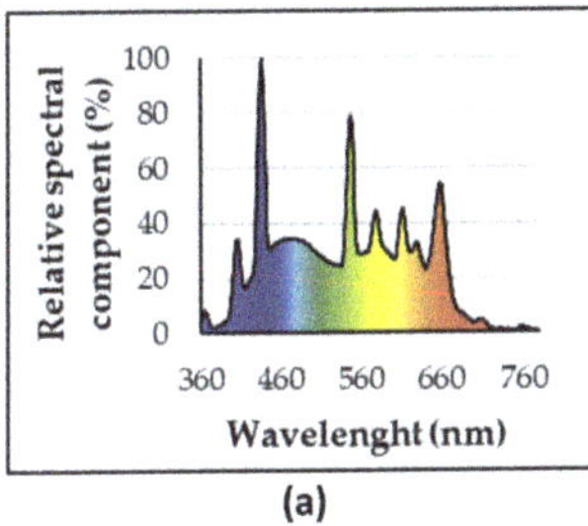
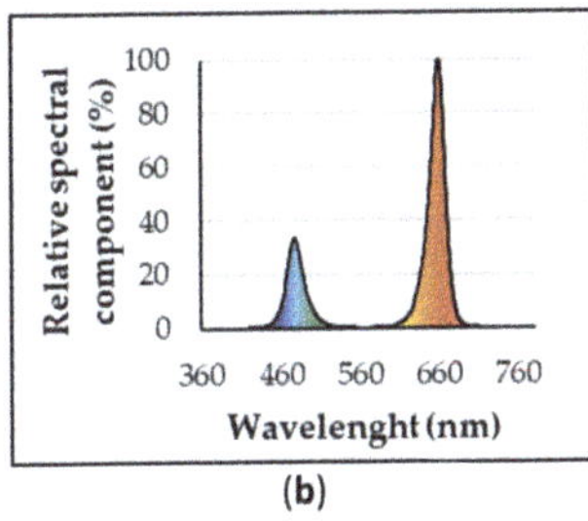
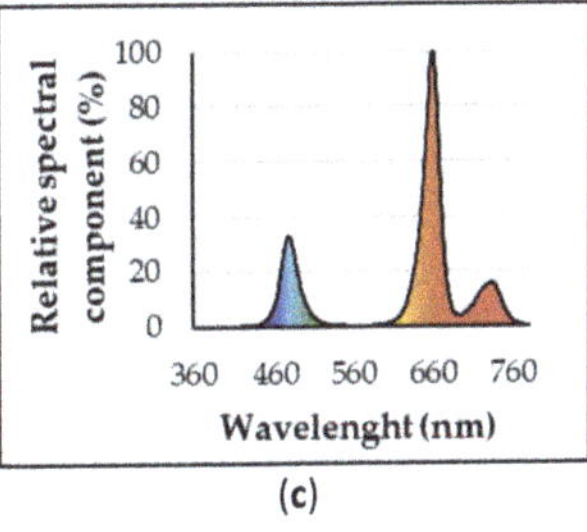

(a)   (b)   (c)

Figure 1. Spectral composition of the light supplied by FL (**a**), RB (**b**) and RB+FR (**c**) treatments used during the experiment.

2.3. Agronomical and Biochemical Parameters

Biomass production corresponding to the shoots weight (yield) was calculated at harvest and expressed as kg m^{-2}. Afterwards, ion content, total phenolic content (TPC), total flavonoid content (TFC), total antioxidant capacity (TAC), total carotenoid content (Car), total chlorophyll content (Chl) and fatty acids were analyzed.

Ion content was analyzed using 0.2 g of dried microgreens and quantified according to [4]. TPC, TFC, TAC, Chl and Car were determined using a methanolic extract, which was obtained from 50 mg of previously lyophilized leaves (Telstar Lyoquest 85 plus, Eco, Barcelona, Spain) in 1.5 mL of methanol. It was then vortexed and incubated overnight at 4 °C. After incubation, it was centrifuged at $16,000\times$ g for 5 min at 4 °C. The supernatant was used as methanolic extract for subsequent analysis. TPC was determined by the Folin-Ciocalteu colorimetric method [22]. TFC was determined as described by Meda et al. [23]. TAC was evaluated in terms of their free radical-scavenging capacity [24]. Chl and Car were measured by extraction of 50 mg of sample in 1.5 mL of MeOH, and then absorbances at 652, 665 and 470 nm were measured in an UV-visible spectrophotometer (8453, Agilent, Santa Clara, CA, USA). The equations developed by Lichtenthaler and Buschmann [25] were used to determine the concentrations of chlorophyll a (Chl a = 16.72 × A665 − 9.16 × A652), chlorophyll b (Chl b = 34.09 × A652 − 15.28 × A665), Chl = Chl a + Chl b, Car = (1000 × A470 − 1.63 × Chl a − 104.96 × Chl b)/221). Chl and Car were expressed as mg kg^{-1} FW. Each of the three replicates was analyzed in triplicate. Fatty acid methyl esters (FAME) were identified and quantified following the methodology of O'Fallon et al. [26], with some modification [27].

2.4. Statistical Analysis

Data were analyzed using Statgraphics Plus. An analysis of variance of agronomical and biochemical parameters (two-way ANOVA) was performed, in which the factors were lighting (FL, RB, RB+FR) and salinity (0 and 80 mM NaCl). When interactions were significant, they were included in the ANOVA; an LSD test was performed to compare lighting treatment and level of salinity.

3. Results and Discussion

3.1. Yield

Yield was affected by light and salinity (Table 1). It increased by 21% in both LED lighting treatments (RB and RB+FR) respect to FL. This finding is consistent with previous results on lettuce and basil, which demonstrated that a mixture of R and B diodes with an RB ratio of 3 resulted in higher yield compared to fluorescent light and other RB ratios [13,21]. This result may be explained by the fact that wavelengths of red and blue lights have been considered as the most convenient sources for enhancing vegetables growth, since they are absorbed by photosynthetic pigments more efficiently than other regions of the light spectrum [28,29]. In our experiment, the addition of FR to the spectrum did not result in a yield increase. However, Jin et al. [30] showed an increase in shoot weight when FR radiation was added to B+R radiation in lettuce 28 days after transplanting and Meng and Runkle [31] in lettuce and basil seedlings, a phenomenon that could be associated with faster leaf area expansion, which would ultimately result in increased light interception [30]. Possibly, in the current experiment, given that microgreens are harvested at early stage, the beneficial effects on leaf elongation induced by FR radiation were negligible, and therefore did not allow for increased light interception.

Table 1. Influence of light treatment and salinity (80 mM NaCl) on yield and nitrate content of purslane microgreens.

	Yield (kg m^{-2})	Nitrate (mg kg^{-1} FW)
Light treatment (A)		
FL	1.47 ± 0.12 a	6392 ± 120 c
RB	1.88 ± 0.16 b	1214 ± 64 b
RB+FR	1.87 ± 0.14 b	556 ± 42 a
Salinity (B)		
0 mM NaCl	1.51 ± 0.09 a	2857 ± 501 b
80 mM NaCl	1.97 ± 0.15 b	2585 ± 529 a
Significant Differences		
A	*	***
B	**	**
A × B	Ns	ns

Asterisk indicates significances at * $p < 0.05$, ** $p < 0.01$, *** $p < 0.001$; ns: non-significant. Different letters in the same column indicate significant differences. FL: fluorescent lamps, RB: LED Red+Blue, RB+FR: LED Red+Blue+Far-Red.

Salinity conditions increased microgreens yield by 23% (Table 1). It has been demonstrated that purslane can be considered as moderately salt tolerant, having the capacity of growing under salt stress conditions [6]. Accordingly, it was previously shown that purslane plants (cv. POR-2936) exposed to 100 mM NaCl presented higher shoot and root productivity, as compared to those grown with 0 mM NaCl [5]. However, the growth of an Aegean purslane accession was more suppressed under 140 mM NaCl than 70 mM NaCl [32], suggesting that the salinity concentration threshold to affect growth differs for each genotype [33]. Thus, increases in fresh weight rising salinity have been recorded in some purslane accessions (e.g., in Ac9), even up to 32 dS m^{-1} [12]. On the other hand, Franco et al. [9] did not find yield differences when the EC in the nutrient solution was increased from 2.5 to 5.0 dS m^{-1} when testing salinity response on purslane accession CM 01-215, while yield was instead reduced by 26% when salinity of 15.0 dS m^{-1} was imposed.

Finally, there was no interaction between light and salinity for yield (Table 1), in agreement with the results of Bantis et al. [34], who did not detect yield differences in spinach baby leaf among mildly saline (40 mM NaCl) and non-saline treatments for the different LEDs tested.

3.2. Anions and Cations

Nitrate content was affected by light and salinity (Table 1), reaching the highest values when microgreens were grown under FL light (6392 mg kg^{-1} FW). This content was reduced by 81% and 91% when they were grown under RB and RB+FR, respectively, regarding FL light. The effects of light spectra on nitrate assimilation in plants are quite complex because the activity of nitrate reductase (NR) and its expression are affected by them. The high value of nitrate content found in microgreens grown under FL light could be due to the lamps used for this treatment featuring an RB ratio of 0.74, much lower than the others (RB ratio of 3). It has been recently demonstrated that red light is effective in reducing nitrate concentration in rocket by increasing NR activity [35], being mediated by phytochrome photoreceptors through transcriptional and posttranslational regulation [36]. However, the effects of blue light on nitrate uptake and utilization in plants seem to be much weaker than those of red light, although when it was combined with other light spectra, it showed positive effects by decreasing nitrate concentration [37]. Thus, Kyriacou et al. [19] demonstrated that combining red and blue light was more effective at promoting nitrate assimilation, resulting in lower nitrate residual concentrations in microgreens. Far-red light addition led us to a higher decrease in nitrate content respect to RB, in agreement with the results of Viršile et al. [38]. A possible explanation for this might be that FR light could induce the phosphorylation of NR through phytochrome-mediated signaling, which would result in its activation [39]. However, the involvement of NR in the response of purslane microgreens to FR and RB needs further investigation, since light effects on nitrate assimilation are plant species-specific [38].

The addition of NaCl reduced nitrate content by 9.5%. This result is consistent with that of Franco et al. [9], who demonstrated that when increasing salinity (>5.0 dS m^{-1}), the nitrate contents of purslane plants was decreased. The difference in nitrate accumulation in response to salinity is generally associated with the inhibition of NO_{3-} uptake by Cl^- [40], which might occur by the interaction between these ions at the site of entrance and for ion transport [41,42]. However, a recent study in basil demonstrates that the interaction between nitrates and chlorines exerts a large effect on yield and metabolomics profile that cannot be satisfactorily explained only by an anion/anion antagonist outcome [43]. They suggest a dose-dependent effect, which is mainly due to the combination of a response to nutrient availability, and an inducible response to stress provoked when chloride concentration in the nutrient solution surpasses that which is accomplished to satisfy nutrient requirements.

There was a significant interaction between light and salinity for oxalate content. The highest content was obtained when microgreens were grown under RB light in non-saline conditions (5311 mg kg^{-1} FW) and the lowest when they were grown under the same light in saline conditions (4375 mg kg^{-1} FW); i.e., saline conditions provoked a reduction in oxalate by 17% (Figure 2). Furthermore, in non-saline conditions, the addition of FR light to RB provoked a significant decrease in oxalate content by 11%. Phytochrome mediates changes in the activities of many enzymes. Particularly, the activity of ascorbate oxidase was reported to increase by irradiation with continuous FR light [44,45], since ascorbic acid accumulation is controlled by the active phytochrome [46]. It has been proposed that oxalate is synthesized via three precursors—glyoxylate/glycolate, ascorbate and oxaloacetate [47]—ascorbic acid being a significant carbon source of oxalic acid in plants. Therefore, it may be advanced that the reduction in oxalate content in purslane microgreens growing under RB+FR lighting as compared with RB under non-saline conditions can be that FR light promotes the activity of ascorbate oxidase, reducing the concentration of ascorbic acid, a major substrate for synthesis of oxalic acid. To date, literature on the influence of light quality on oxalate content is limited. Qi et al. [48] applied four lighting treatments using FL lamps of different colors (red, blue, yellow and white) over a soilless crop of spinach, showing that oxalate content in spinach leaves under red light was 47.6% lower than that under white light. Gao et al. [49], in another study on hydroponic spinach using LED lamps with different ratio of red light to blue light (0.9, 1.2, and 2.2), concluded that treatment with more fraction of red light was favorable to reduce oxalate accumulation. These results, although elaborated with different crop species and light sources from the hereby presented experiment, substantiate the need for additional research on the role of light in oxalate content and ascorbic acid pathways with possible applications also for purslane.

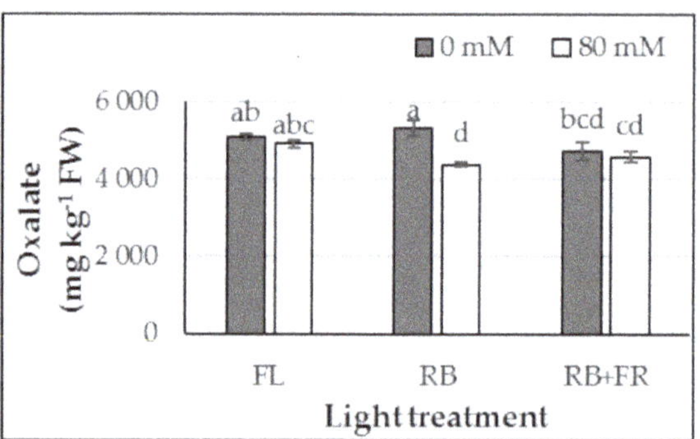

Figure 2. Effect of the different light (FL, RB and RB+FR) and salinity (0 and 80 mM NaCl) combinations on the oxalate content of purslane microgreens at harvest. Values are the mean $\pm$ SE (n = 9). Different letters indicate significant differences ($p \leq 0.05$).

The decrease in oxalate by adding NaCl to the nutrient solution was significant only when RB lighting was used. In previous studies, [50] suggested that when chloride or other anions are absorbed by plants, these anions compete for cations and depress oxalate synthesis. The decrease in oxalic acid accumulation on the leaves could therefore be associated to a competitive accumulation between oxalic acid with chloride ion, since purslane plant accumulates chloride ions in the leaves when submitted to saline stress [6]. Oxalate accumulation in vacuoles plays a vital function in balancing excess of inorganic cations over anions [33], regulating intracellular pH to neutralize excess cations [51]. Oxalate concentration was not affected by saline conditions in the other lighting treatments, which could be due to the influence of different spectra, which could affect the balance between cations and anions and/or the activity of ascorbate oxidase as previously mentioned in the case of FR lighting, although other mechanisms may operate; therefore, further research is needed.

There was a significant interaction between light and salinity for Cl, Na, Ca and K content. Cl and Na content had the same pattern among different treatments, with the highest contents being found when microgreens were grown under FL+NaCl treatment and the lowest under RB and RB+FR in non-saline conditions (Figure 3a,b). Accordingly, both in saline and non-saline conditions, Cl and Na contents of microgreens grown under LED lights were significantly reduced as compared to those grown under FL light. Therefore, spectral quality could modulate responses to a salinity stress probably through regulation of antioxidant enzymes and non-enzymatic systems, with phytochrome B1 playing a very important role in this process [52], but further research is needed for confirming this. As it was expected, the addition of NaCl to the nutrient solution significantly increased the Cl and Na content in each light treatment.

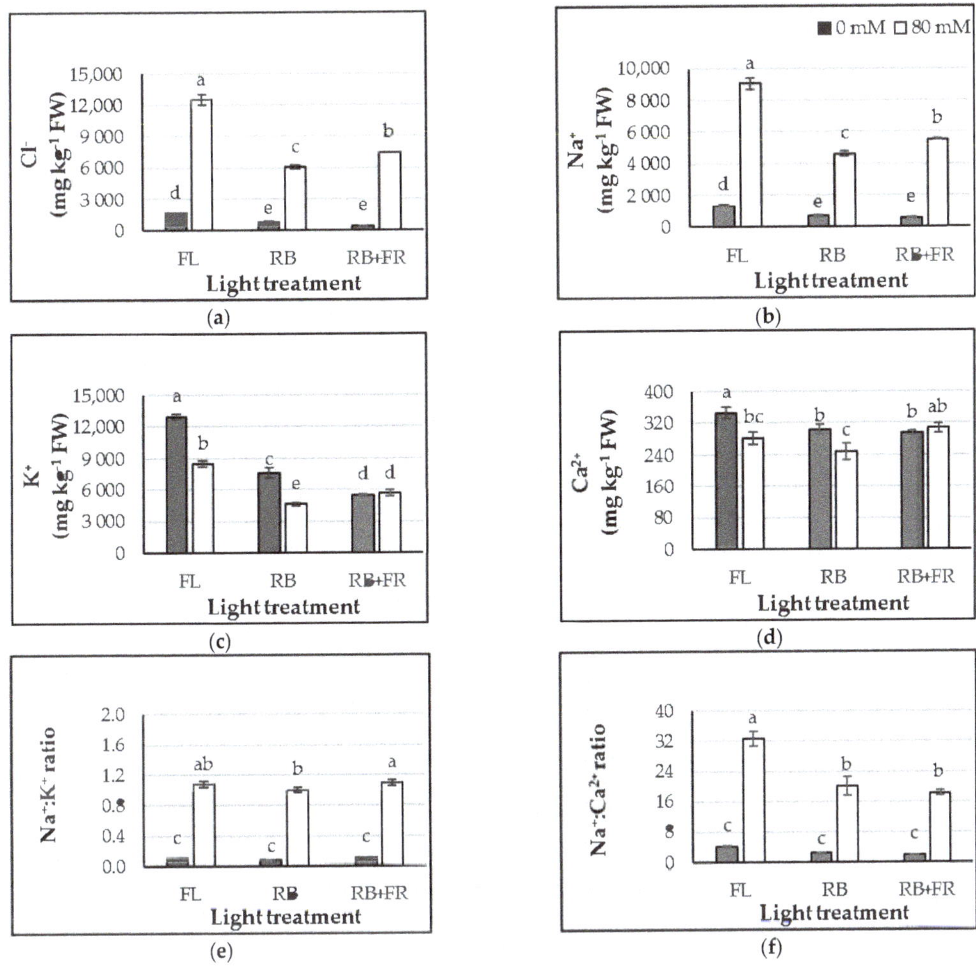

Figure 3. Effect of the different light (FL, RB and RB+FR) and salinity (0 and 80 mM NaCl) combinations on Cl⁻ (**a**), Na⁺ (**b**), K⁺ (**c**), and Ca²⁺ (**d**) content and Na⁺/K⁺(**e**) and Na⁺/Ca²⁺ (**f**) ratios of purslane microgreens at harvest. Values are the mean ± SE (n = 9). Different letters indicate significant differences ($p \leq 0.05$).

With respect to K content (Figure 3c), the lowest values (4601 mg kg⁻¹ FW) were found when the microgreens were grown under RB light in saline conditions, while the highest values (12,970 mg kg⁻¹ FW) were obtained when they were grown under FL light in non-saline conditions. Ca content had a similar pattern than K (Figure 3d), but there were no significant differences for lowest Ca and for the highest Ca values between RB and FL lights in saline and non-saline conditions, respectively. Furthermore, the addition of FR to the spectrum increased K and Ca contents in saline condition. However, salinity reduced their contents under RB and FL lights. The Na:K ratio was about 1 for saline conditions and about 0.1 for non-saline conditions, independently of the light treatment (Figure 3e). However, the Na:Ca ratio ranged between 32 for FL light and around 18 for LED lights in saline conditions, and between 4 and 2 for non-saline conditions for all lights (Figure 3f). Therefore, salinity could induce both osmotic stress and ionic toxicity associated with excessive Cl and Na uptake, leading to Ca and K shortage and to other nutrient imbalances [53], as was previously demonstrated in purslane [54]. Essentially, the K decrease in saline conditions can be explained by the antagonism of Na and K at uptake places in the roots, the effect of Na on K transport into the xylem or the inhibition of the uptake process [55]. Regarding the Ca decrease, it could be due not only to the accumulation of Na in the leaves, but also because of its reduced mobility and transport in plant due to the lower water uptake [56,57]. In summary, the addition of FR light led to a higher increase in Cl, Na, K and Ca content in leaves compared to RB light in saline conditions, maintaining the ratio Na/Ca and increasing the ratio Na/K.

3.3. Total Phenol Content, Total Flavonoids Content and Total Antioxidant Capacity

TPC was only affected by lighting, with the microgreens grown under the RB spectrum being those that reached the highest content (Table 2). Martínez-Zamora et al. [58] found that the TPC of carrot sprouts remained quite constant during a germination period of 7 d in darkness plus 10 d in a similar photoperiod in samples grown under RB LEDs, while Fl and RB+FR treatments reduced 24% and 12% the maximum TPC reached during growing, respectively. Purslane microgreens under RB+FR treatment showed the lowest TFC (around 14% lower than under FL and RB lights), while salinity reduced the TFC by 10%. In addition, the TAC was not affected by light or salinity treatments. However, Martínez-Zamora et al. [58] reported that under the above lighting conditions, carrot sprouts from the FL treatment reported 39.4% less TAC, by the same procedure used in this experiment, than those grown under RB LEDs. In fact, in a review concerning LED lighting as a key to modulate antioxidant compounds in plants, Loi et al. [59] stated that B, R and FR lights had shown to be able to increase the TPC and TFC in different plant commodities, and that the regulation of TPC and TFC by LEDs can be performed directly by inducing the expression of the key enzyme and indirectly by increasing the shikimic acid as precursor molecule [60]. Nevertheless, besides light intensity, their contents also depend upon plant species, cultivars and timing of LED exposure [59]. It is well established that salinity can enhance the synthesis of secondary metabolites in plants [61], producing reactive oxygen species, which are harmful to plant cells. Thus, polyphenols could help in improving TAC of plants at low to moderate salt concentrations [62]. Therefore, plants such as halophytes tolerant to stress could be interesting organisms for production of secondary metabolites, which would be useful for food and medicinal applications [63]. In our experiment, in contrast, TPC and TAC was not affected by salinity, and TFC was decreased when microgreens were grown with 80 mM NaCl (Table 2). However, in agreement with our results, He et al. [5] revealed that purslane grown with 0 and 100 mM NaCl had similar TPC, which decreased when plants were grown with 200 and 300 mM NaCl. According to the previous authors, TPC for purslane grown under higher salinity could be associated with their lower photosynthetic performance, although genetics and environment can be also affecting their accumulation [64]. It has also been demonstrated that purslane plants responded to NaCl stress by enhancing their antioxidative capacity and proline accumulation [32]. However, TAC is regulated by the level of stress associated to NaCl concentration. In our experiment, TAC was not affected by salinity, indicating an overtaking of antioxidant capacity face to salt stress, as could in the halophyte *Enchylaena tomentosa* at high salinity conditions [65]. Similar results were reported by Islam et al. [66], where low and medium levels of NaCl increased the TAC in wheat microgreens, whereas high salt-concentration (100 mM) suppressed the accumulation of antioxidants production, and by Valifard et al. [67], who showed that there was an increase and decrease in antioxidants in *Salvia mirzayanii* leaves because of mild and high salinity stress, respectively. Furthermore, in our experiment, salt stress decreased the TFC, in agreement with Islam et al. [66], who demonstrated that the production of an extract from wheat microgreen with above the concentration of 50 mM Na Cl resulted in a decrease in TFC, and with Zhou et al. [68], who showed that the TFC increased at low (25 mM) or moderate (50 mM) levels but declined at severe (75 and 100 mM) levels in *Schizonepeta tenuifolia*. It has been hypothesized that the biosynthesis of flavonoids can be encouraged by the changes in the cellular redox homeostasis, which are regulated by changing cellular redox potentials [69,70].

Table 2. Influence of light and salinity (80 mM NaCl) on the total phenolics, total flavonoids and antioxidant capacity of purslane microgreens.

	Total Phenolics (mg GA kg^{-1} FW)	Total Flavonoids (mg Rutin kg^{-1} FW)	Antioxidant Capacity (mg DPPH$_{reduced}$ kg^{-1} FW)
Treatments (A)			
FL	1176 ± 21 b	2829 ± 72 b	170 ± 3
RB	1277 ± 27 c	2819 ± 91 b	174 ± 5
RB+FR	931 ± 25 a	2417 ± 91 a	177 ± 6
Salinity (B)			
0 mM NaCl	1112 ± 64	2834 ± 99 b	175 ± 6
80 mM NaCl	1143 ± 69	2544 ± 98 a	172 ± 3
Significant Differences			
A	***	***	ns
B	ns	**	ns
A × B	ns	ns	ns

Asterisk indicates significances at ** $p < 0.01$, *** $p < 0.001$, ns: non-significant. Different letters in the same column indicate significant differences. GA: gallic acid; DPPH: 2.2-diphenyl-1-picrythydrazyl. (FL: fluorescent lamps, RB: LED Red+Blue, RB+FR: LED Red+Blue+Far-Red).

3.4. Chlorophylls and Carotenoids

Regarding photosynthetic pigments (Car and Chl), a significant interaction between light and salinity factors was observed. The highest values for Car were obtained under LED lighting in non-saline conditions, whereas the lowest values were achieved in salinity conditions, without differences among factor combinations (Figure 4a). Regarding Chl, there were no differences between lighting treatments in non-saline conditions, while salinity conditions reduced their contents for every factor combination, reaching the lowest value under RB+FR treatment (Figure 4b). It has been demonstrated that light quantity and quality are the most important factors affecting chlorophyll and carotenoid contents because they are closely related to photosynthesis. In this context, blue LEDs, used alone or in combination with red light, increased the carotenoid and chlorophyll content [71,72]. Similarly, Samouliene et al. [73] found that higher concentrations of carotenoids and chlorophylls a and b were found when B light component was increased by 33% in comparison to lower B light dosages. Furthermore, Li and Kubota [71] demonstrated that supplemental FR decreased carotenoids and chlorophyll concentration by 11% and 14%. In our experiment, only Car was affected by light in non-saline conditions, increasing their concentrations under RB and RB+FR by 29% and 26%, respectively, in relation to FL light, possibly due to their higher percentage in B light of the two former lights. In contrast, adding FR to RB light did not affect either Car or Chl in non-saline conditions, but decreased Chl in saline conditions.

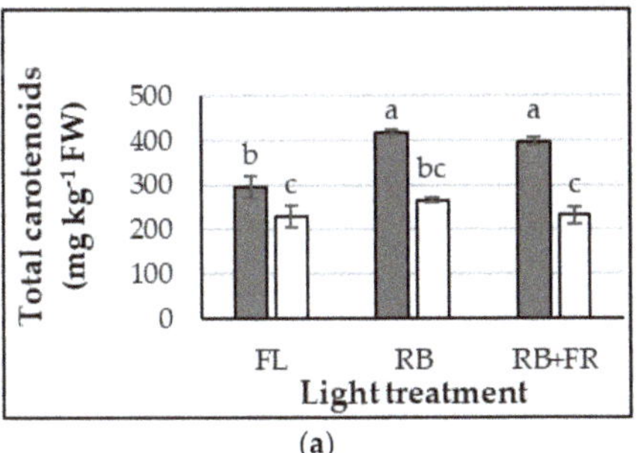

(a)

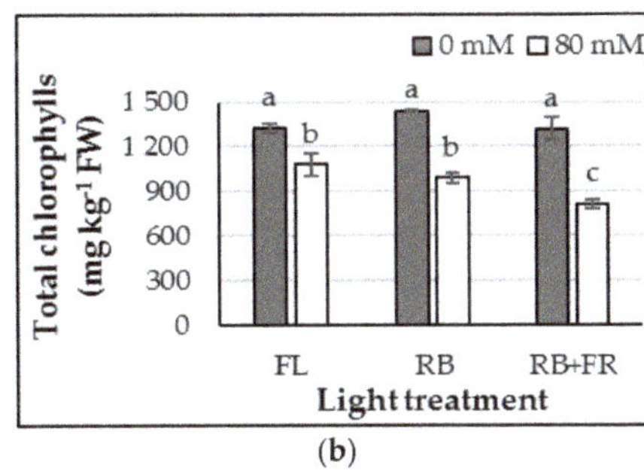

(b)

Figure 4. Effect of the different light (FL, RB and RB+FR) and salinity (0 and 80 mM NaCl) combinations on total carotenoids (**a**) and total chlorophylls (**b**) of purslane microgreens. Values are mean ± SE (n = 3). Different letters indicate significant differences ($p \leq 0.05$).

Car and Chl in microgreens were reduced by salinity in the three lighting treatments. Camelle et al. [33] found a significant effect of the salinity, the nitrate to ammonium ratio and their interaction on Car and TChl accumulation in two purslane ecotypes, with salinity in plants growing with the 25:75 nitrate:ammonium ratio, leading to a drop in Car and Chl. A decrease in both parameters were also observed in the ecotype ET growing with the 66:33 ratio. He et al. [5] showed for total Car, that purslane plants grown with 100 mM NaCl reached the highest value followed by those grown with 0 mM, then 200 mM and later 300 mM NaCl. In the case of Chl, they did not find differences in Chl between purslane grown with 0 and with 100 mM NaCl, but they were higher than those grown with 200 and 300 mM NaCl. In addition, the decrease in the Chl of purslane plants with an increase in salinity in the nutrient solution has been previously demonstrated by Franco et al. [9], in agreement with our results. Finally, Zaman et al. [74] found that an increase in NaCl concentration from 0 to 200 mM caused a decrease in Chl in two genotypes of purslane. The results discussed here allow us to suppose that salinity stress could affect the photosynthetic performance in purslane due to salt accumulation in their leaves [75] and the decreases in Car and Chl, which can be altered according to level of salinity, the nitrate:ammonium ratio and the genotypes used.

3.5. Fatty Acids

The fatty acid content ranged between 303 and 418 mg 100 g^{-1} FW, without significant differences between treatments (Supplementary Table S1). The major fatty acid was α-linolenic acid (LNA), whose proportion with respect to total fatty acids ranged between 54% and 60%, followed by linoleic acid (LA) (between 24% and 26%), palmitic acid (between 7% and 8%) and oleic acid (between 3% and 63%) (Figure 5). Smaller amounts of lauric acid, stearic acid, myristic acid, arachidic acid, behenic acid and lignoceric acid were also detected, although the latter was not detected in microgreens grown under RB+FR in both saline and non-saline conditions. As a consequence, oleic acid reached the highest content in these treatments. The LNA:LA ratio was significantly higher in the microgreens grown under FL in both saline and non-saline conditions, and under RB and RB+FR in non-saline conditions, compared with the ratio obtained under RB and RB+FR in saline conditions. To our knowledge, no study has been conducted on the effect of LED lighting on fatty acid content in superior plants, with most of the studies dealing with major of them having been conducted on microalgae. In these species, it has been demonstrated that the expression of fatty acid biosynthesis genes is regulated by different LEDs and that, consequently, different light intensities and wavelengths affect their fatty acid yields [76]. In our study, the change in fatty acid composition under RB+FR light increased the oleic acid content. In contrast, Sharma et al. [77] found that specifically polyunsaturated FAs were augmented, whereas monounsaturated FAs decreased in *Phaeodactylum tricornutum* grown under red light compared to white light. Future studies on the current topic are therefore recommended, using different spectra composition, in order to get a convenient fatty acid composition in purslane.

As regards to salinity, our results agree with those obtained by Carvalho et al. [10], who found that the total amount of fatty acids in purslane did not significantly change with increasing saline stress. Although the unsaturated fatty acids have emerged as general protectors against different abiotic and biotic stress [78], in our study, this effect was not observed, probably due to the salinity tolerance of purslane microgreens. However, Camalle et al. [33] observed differences with respect to total fatty acid content for the purslane ecotypes and nitrate to ammonium ratio tested, with an increase under saline conditions for ecotype ET compared to ecotype RN. In addition, changes in the fatty acid composition of leaves were detected and fatty acid contents were affected by salinity in borage, with LNA content being the most affected [79]. Therefore, besides salinity, differences in fatty acid content may be the result of differences in plant ecotypes or environmental and developmental factors, plant growth stage, nutrient solution composition, sample material or sampling procedures [33,80].

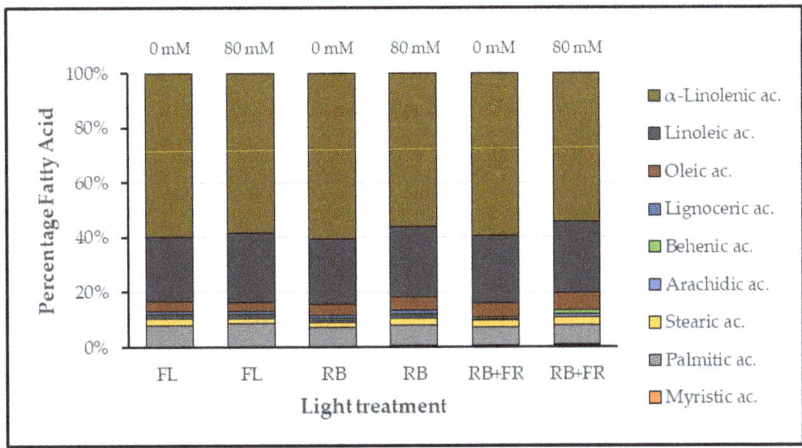

Figure 5. Percentage of fatty acid content of the aerial part of purslane microgreens for the different light treatment (FL, RB and RB+FR) and salinity (0 and 80 mM NaCl) combinations.

4. Conclusions

The results of this study indicate that it is convenient to grow purslane microgreens with 80 mM NaCl under RB and RB+FR lights, since these conditions improve yield and quality of the product, reducing the content of anti-nutritional compounds. In addition, the concentration of Cl and Na in the leaves can be reduced using the above conditions. However, a salinity of 80 mM reduced the content of total flavonoids, chlorophylls and carotenoids. Total fatty acids were not affected by salinity or lighting, but the specific composition of them can be modified by these factors. However, more research using different salinity conditions and spectra and intensity of LED lights is needed to target specific improvements of microgreens purslane quality. In further experiments, quality changes during shelf life should be elucidated.

Supplementary Materials: The following are available online at https://www.mdpi.com/article/10.3390/horticulturae7070180/s1. Table S1: title, Fatty acids (mg gr^{-1} FW) of purslane microgreens grown under the different light treatments (FL: fluorescent lamps, RB: LED Red Blue, RB+FR: LED Red Blue+Far-Red) and salinity (0, 80 mM NaCl) (n = 3).

Author Contributions: Conceptualization, M.d.C.M.-B. and J.A.F.; methodology, G.P., F.O. and A.C.; formal analysis, A.G.; investigation, A.G. and P.A.G.; resources, A.C. and J.A.F.; data curation, A.G. and F.O.; writing—original draft preparation, A.G., M.d.C.M.-B. and J.A.F.; writing—review and editing C.E.-G., F.A.-H., F.O., G.P., M.d.C.M.-B. and J.A.F.; supervision, F.O., C.E.-G., P.A.G. and F.A.-H.; funding acquisition, J.A.F. All authors have read and agreed to the published version of the manuscript.

Funding: This project was financed by the Autonomous Community of the Region of Murcia n° 20849/PI/18 through the grant call for projects for the development of scientific and technical research by competitive groups, included in the Regional Programme for the Promotion of Scientific and Technical Research (Action Plan 2018) of the Seneca Foundation-Science and Technology Agency of the Region of Murcia (Spain). The research also received funding from the Italian Ministry of Agricultural, Food and Forestry Policies (MIPAAF), in the framework of the project "Light on Shelf Life" (CUP J56J20000410008).

Institutional Review Board Statement: Not applicable.

Informed Consent Statement: Not applicable.

Acknowledgments: The elaboration of the manuscript was supported by a grant from the Fundación Séneca (reference 20555/IV/18, Call for Fellowships for Guest Researcher Stays at Universities and OPIS of the Region of Murcia) awarded to Francesco Orsini.

Conflicts of Interest: The authors declare no conflict of interest.

References

1. Cros, V.; Martínez-Sánchez, J.J.; Franco, J.A. Good Yields of Common Purslane with a High Fatty Acid Content Can Be Obtained in a Peat-based Floating System. *HortTechnology* **2007**, *17*, 14–20. [CrossRef]
2. Egea-Gilabert, C.; Ruiz-Hernández, M.V.; Parra, M.Á.; Fernández, J.A. Characterization of purslane (*Portulaca oleracea* L.) accessions: Suitability as ready-to-eat product. *Sci. Hortic.* **2014**, *172*, 73–81. [CrossRef]
3. Gonnella, M.; Charfeddine, M.; Conversa, G.; Santamaria, P. Purslane: Review of its Potential for Health and Agricultural Aspect. The European. *J. Sci. Biotech.* **2010**, *4*, 131–136.
4. Lara, L.J.; Egea-Gilabert, C.; Niñirola, D.; Conesa, E.; Fernández, J.A. Effect of aeration of the nutrient solution on the growth and quality of purslane (*Portulaca oleracea*). *J. Hortic. Sci. Biotechnol.* **2011**, *86*, 603–610. [CrossRef]
5. He, J.; You, X.; Qin, L. High Salinity Reduces Plant Growth and Photosynthetic Performance but Enhances Certain Nutritional Quality of C4 Halophyte *Portulaca oleracea* L. Grown Hydroponically Under LED Lighting. *Front. Plant Sci.* **2021**, *12*, 651341. [CrossRef] [PubMed]
6. Teixeira, M.; Carvalho, I.S. Effects of salt stress on purslane (*Portulaca oleracea*) nutrition. *Ann. Appl. Biol.* **2008**, *154*, 77–86. [CrossRef]
7. Karakaş, S.; Çullu, M.A.; Dikilitaş, M. Comparison of two halophyte species (*Salsola soda* and *Portulaca oleracea*)for salt removal potential under different soil salinity conditions. *Turk. J. Agric. For.* **2017**, *41*, 183–190. [CrossRef]
8. Alam, A.; Juraimi, A.; Rafii, M.; Hamid, A.; Aslani, F.; Alam, M. Effects of salinity and salinity-induced augmented bioactive compounds in purslane (*Portulaca oleracea* L.) for possible economical use. *Food Chem.* **2015**, *169*, 439–447. [CrossRef] [PubMed]
9. Franco, J.A.; Cros, V.; Vicente, M.J.; Martínez-Sánchez, J.J. Effects of salinity on the germination, growth, and nitrate contents of purslane (*Portulaca oleracea* L.) cultivated under different climatic conditions. *J. Hortic. Sci. Biotechnol.* **2011**, *86*, 1–6. [CrossRef]
10. Carvalho, I.S.; Teixeira, M.; Brodelius, M. Effect of Salt Stress on Purslane and Potential Health Benefits: Oxalic Acid and Fatty Acids Profile. 2009. Available online: https://escholarship.org/uc/item/4cc78714 (accessed on 20 April 2021).
11. Anastácio, A.; Carvalho, I.S. Accumulation of fatty acids in purslane grown in hydroponic salt stress conditions. *Int. J. Food Sci. Nutr.* **2012**, *64*, 235–242. [CrossRef]
12. Alam, A.; Juraimi, A.S.; Rafii, M.Y.; Hamid, A.A.; Aslani, F.; Hakim, M.A. Salinity-induced changes in the morphology and major mineral nutrient composition of purslane (*Portulaca oleracea* L.) accessions. *Biol. Res.* **2016**, *49*, 24. [CrossRef]
13. Pennisi, G.; Orsini, F.; Blasioli, S.; Cellini, A.; Crepaldi, A.; Braschi, I.; Spinelli, F.; Nicola, S.; Fernandez, J.A.; Stanghellini, C.; et al. Resource use efficiency of indoor lettuce (*Lactuca sativa* L.) cultivation as affected by red:blue ratio provided by LED lighting. *Sci. Rep.* **2019**, *9*, 14127. [CrossRef]
14. Castillejo, N.; Martínez-Zamora, L.; Gómez, P.A.; Pennisi, G.; Crepaldi, A.; Fernández, J.A.; Orsini, F.; Artés-Hernández, F. Postharvest LED lighting: Effect of red, blue and far red on quality of minimally processed broccoli sprouts. *J. Sci. Food Agric.* **2020**, *101*, 44–53. [CrossRef] [PubMed]
15. Pennisi, G.; Orsini, F.; Castillejo, N.; Gómez, P.A.; Crepaldi, A.; Fernández, J.A.; Egea-Gilabert, C.; Artés-Hernández, F.; Gianquinto, G. Spectral composition from led lighting during storage affects nutraceuticals and safety attributes of fresh-cut red chard (*Beta vulgaris*) and rocket (*Diplotaxis tenuifolia*) leaves. *Postharvest Biol. Technol.* **2021**, *175*, 111500. [CrossRef]
16. Rouphael, Y.; Kyriacou, M.; Petropoulos, S.A.; De Pascale, S.; Colla, G. Improving vegetable quality in controlled environments. *Sci. Hortic.* **2018**, *234*, 275–289. [CrossRef]
17. Samuolienė, G.; Brazaitytė, A.; Viršilė, A.; Miliauskienė, J.; Vaštakaitė-Kairienė, V.; Duchovskis, P. Nutrient Levels in Brassicaceae Microgreens Increase Under Tailored Light-Emitting Diode Spectra. *Front. Plant Sci.* **2019**, *10*, 1475. [CrossRef]
18. Alrifai, O.; Hao, X.; Marcone, M.F.; Tsao, R. Current Review of the Modulatory Effects of LED Lights on Photosynthesis of Secondary Metabolites and Future Perspectives of Microgreen Vegetables. *J. Agric. Food Chem.* **2019**, *67*, 6075–6090. [CrossRef]
19. Kyriacou, M.C.; El-Nakhel, C.; Pannico, A.; Graziani, G.; Soteriou, G.A.; Giordano, M.; Zarrelli, A.; Ritieni, A.; De Pascale, S.; Rouphael, Y. Genotype-Specific Modulatory Effects of Select Spectral Bandwidths on the Nutritive and Phytochemical Composition of Microgreens. *Front. Plant Sci.* **2019**, *10*, 1501. [CrossRef]
20. Egea-Gilabert, C.; Fernández, J.A.; Migliaro, D.; Martínez-Sánchez, J.J.; Vicente, M.J. Genetic variability in wild vs. cultivated Eruca vesicaria populations as assessed by morphological, agronomical and molecular analyses. *Sci. Hortic.* **2009**, *121*, 260–266. [CrossRef]
21. Pennisi, G.; Sanyé-Mengual, E.; Orsini, F.; Crepaldi, A.; Nicola, S.; Ochoa, J.; Fernandez, J.; Gianquinto, G. Modelling Environmental Burdens of Indoor-Grown Vegetables and Herbs as Affected by Red and Blue LED Lighting. *Sustainability* **2019**, *11*, 4063. [CrossRef]
22. Everette, J.D.; Bryant, Q.M.; Green, A.M.; Abbey, Y.A.; Wangila, G.W.; Walker, R.B. Thorough Study of Reactivity of Various Compound Classes toward the Folin−Ciocalteu Reagent. *J. Agric. Food Chem.* **2010**, *58*, 8139–8144. [CrossRef]

23. Meda, A.; Lamien, C.E.; Romito, M.; Millogo, J.; Nacoulma, O.G. Determination of the total phenolic, flavonoid and proline contents in Burkina Fasan honey, as well as their radical scavenging activity. *Food Chem.* **2005**, *91*, 571–577. [CrossRef]
24. Brand-Williams, W.; Cuvelier, M.E.; Berset, C. Use of a free radical method to evaluate antioxidant activity. *LWT Food Sci. Technol.* **1995**, *28*, 25–30. [CrossRef]
25. Lichtenthaler, H.K.; Buschmann, C. Chlorophylls and Carotenoids: Measurement and Characterization by UV-VIS Spectroscopy. *Curr. Protoc. Food Anal. Chem.* **2001**, *1*, F4.3.1–F4.3.8. [CrossRef]
26. O'Fallon, J.V.; Busboom, J.R.; Nelson, M.L.; Gaskins, C.T. A direct method for fatty acid methyl ester synthesis: Application to wet meat tissues, oils, and feedstuffs. *J. Anim. Sci.* **2007**, *85*, 1511–1521. [CrossRef] [PubMed]
27. Labiad, M.; Giménez, A.; Varol, H.; Tüzel, Y.; Egea-Gilabert, C.; Fernández, J.; Martínez-Ballesta, M. Effect of Exogenously Applied Methyl Jasmonate on Yield and Quality of Salt-Stressed Hydroponically Grown Sea Fennel (*Crithmum maritimum* L.). *Agronomy* **2021**, *11*, 1083. [CrossRef]
28. Kopsell, D.A.; Sams, C.E.; Barickman, T.C.; Morrow, R.C. Sprouting Broccoli Accumulate Higher Concentrations of Nutritionally Important Metabolites under Narrow-band Light-emitting Diode Lighting. *J. Am. Soc. Hortic. Sci.* **2014**, *139*, 469–477. [CrossRef]
29. Lee, J.S.; Kim, Y.H. Growth and Anthocyanins of Lettuce Grown under Red or Blue Light-emitting Diodes with Distinct Peak Wavelength. *Korean J. Hortic. Sci. Technol.* **2014**, *32*, 330–339. [CrossRef]
30. Jin, W.; Urbina, J.L.; Heuvelink, E.; Marcelis, L.F.M. Adding Far-Red to Red-Blue Light-Emitting Diode Light Promotes Yield of Lettuce at Different Planting Densities. *Front. Plant Sci.* **2021**, *11*, 609977. [CrossRef] [PubMed]
31. Meng, Q.; Runkle, E.S. Far-red radiation interacts with relative and absolute blue and red photon flux densities to regulate growth, morphology, and pigmentation of lettuce and basil seedlings. *Sci. Hortic.* **2019**, *255*, 269–280. [CrossRef]
32. Yazici, I.; Türkan, I.; Sekmen, A.H.; Demiral, T. Salinity tolerance of purslane (*Portulaca oleracea* L.) is achieved by enhanced antioxidative system, lower level of lipid peroxidation and proline accumulation. *Environ. Exp. Bot.* **2007**, *61*, 49–57. [CrossRef]
33. Camalle, M.; Standing, D.; Jitan, M.; Muhaisen, R.; Bader, N.; Bsoul, M.; Ventura, Y.; Soltabayeva, A.; Sagi, M. Effect of Salinity and Nitrogen Sources on the Leaf Quality, Biomass, and Metabolic Responses of Two Ecotypes of Portulaca oleracea. *Agronomy* **2020**, *10*, 656. [CrossRef]
34. Bantis, F.; Fotelli, M.; Ilić, Z.; Koukounaras, A. Physiological and Phytochemical Responses of Spinach Baby Leaves Grown in a PFAL System with LEDs and Saline Nutrient Solution. *Agriculture* **2020**, *10*, 574. [CrossRef]
35. Signore, A.; Bell, L.; Santamaria, P.; Wagstaff, C.; Van Labeke, M.-C. Red Light Is Effective in Reducing Nitrate Concentration in Rocket by Increasing Nitrate Reductase Activity and Contributes to Increased Total Glucosinolates Content. *Front. Plant Sci.* **2020**, *11*, 604. [CrossRef]
36. Sakuraba, Y.; Yanagisawa, S. Light signalling-induced regulation of nutrient acquisition and utilisation in plants. *Semin. Cell Dev. Biol.* **2018**, *83*, 123–132. [CrossRef] [PubMed]
37. Bian, Z.-H.; Lei, B.; Cheng, R.-F.; Wang, Y.; Li, T.; Yang, Q.-C. Selenium distribution and nitrate metabolism in hydroponic lettuce (*Lactuca sativa* L.): Effects of selenium forms and light spectra. *J. Integr. Agric.* **2020**, *19*, 133–144. [CrossRef]
38. Viršilė, A.; Brazaitytė, A.; Sakalauskienė, S.; Jankauskienė, J.; Miliauskienė, J.; Vaštakaitė, V. LED lighting for reduced nitrate contents in green vegetables. *Acta Hortic.* **2018**, *1227*, 669–676. [CrossRef]
39. Chandok, M.R.; Sopory, S.K. Phosphorylation/dephosphorylation steps are key events in the phytochrome-mediated enhancement of nitrate reductase mRNA levels and enzyme activity in maize. *Mol. Gen. Genet.* **1996**, *251*, 599–608. [CrossRef]
40. Aslam, M.; Huffaker, R.C.; Rains, W.D. Early effects of salinity on nitrate assimilation in barley seedlings. *Plant Physiol.* **1984**, *76*, 321–325. [CrossRef] [PubMed]
41. Rubinigg, M.; Posthumus, F.; Ferschke, M.; Elzenga, J.M.; Stulen, I. Effects of NaCl salinity on 15N-nitrate fluxes and specific root length in the halophyte *Plantago maritima* L. *Plant Soil* **2003**, *250*, 201–213. [CrossRef]
42. Varma, S.; Rogers, D.M.; Pratt, L.; Rempe, S.B. Design principles for K^+ selectivity in membrane transport. *J. Gen. Physiol.* **2011**, *137*, 479–488. [CrossRef]
43. Corrado, G.; Lucini, L.; Miras-Moreno, B.; Chiaiese, P.; Colla, G.; De Pascale, S.; Rouphael, Y. Metabolic Insights into the Anion-Anion Antagonism in Sweet Basil: Effects of Different Nitrate/Chloride Ratios in the Nutrient Solution. *Int. J. Mol. Sci.* **2020**, *21*, 2482. [CrossRef]
44. Mapson, L.W. Metabolism of Ascorbic Acid in Plants: Part I. Function. *Annu. Rev. Plant Physiol.* **1958**, *9*, 119–150. [CrossRef]
45. Hayashi, R.; Morohashi, Y. Phytochrome Control of the Development of Ascorbate Oxidase Activity in Mustard (*Sinapis alba* L.) Cotyledons. *Plant Physiol.* **1993**, *102*, 1237–1241. [CrossRef]
46. Bienger, I.; Schopfer, P. Photomodulation by phytochrome of the rate of accumulation of ascorbic acid in mustard seedlings (*Sinapis alba* L.). *Planta* **1970**, *93*, 152–159. [CrossRef] [PubMed]
47. Cai, X.; Ge, C.; Xu, C.; Wang, X.; Wang, S.; Wang, Q. Expression Analysis of Oxalate Metabolic Pathway Genes Reveals Oxalate Regulation Patterns in Spinach. *Molecules* **2018**, *23*, 1286. [CrossRef]
48. Qi, L.; Shiqi, L.; Li, X.; Wenyan, Y.; Liang Qingling, L.; Shuqin, H. Effects of light qualities on accumulation of oxalate, tannin and nitrate in spinach. *Trans. CSAE* **2007**, *23*, 201–205. [CrossRef]
49. Gao, W.; He, D.; Ji, F.; Zhang, S.; Zheng, J. Effects of Daily Light Integral and LED Spectrum on Growth and Nutritional Quality of Hydroponic Spinach. *Agronomy* **2020**, *10*, 1082. [CrossRef]

50. Singh, K.; Chatrath, R. Salinity tolerance. In *Application of Physiology in Wheat Breeding*; Reynolds, M.P., Ortiz-Monasterio, J.I., McNab, A., Eds.; CIMMYT Crop Improvement Division: México DF, Mexico, 2001; Chapter 8; pp. 101–110.

51. Yang, C.; Chong, J.; Li, C.; Kim, C.; Shi, D.; Wang, D. Osmotic adjustment and ion balance traits of an alkali resistant halophyte Kochia sieversiana during adaptation to salt and alkali conditions. *Plant Soil* **2007**, *294*, 263–276. [CrossRef]

52. Cao, K.; Yu, J.; Xu, D.; Ai, K.; Bao, E.; Zou, Z. Exposure to lower red to far-red light ratios improve tomato tolerance to salt stress. *BMC Plant Biol.* **2018**, *18*, 92. [CrossRef] [PubMed]

53. Marschner, H. *Mineral Nutrition of Higher Plants*, 2nd ed.; Academic Press: London, UK, 1995.

54. Kafi, M.; Rahimi, Z. Effect of salinity and silicon on root characteristics, growth, water status, proline content and ion accumulation of purslane (*Portulaca oleracea* L.). *Soil Sci. Plant Nutr.* **2011**, *57*, 341–347. [CrossRef]

55. Hu, Y.; Schmidhalter, U. Drought and salinity: A comparison of their effects on mineral nutrition of plants. *J. Plant Nutr. Soil Sci.* **2005**, *168*, 541–549. [CrossRef]

56. Coskun, D.; Britto, D.T.; Jean, Y.-K.; Kabir, I.; Tolay, I.; Torun, A.A.; Kronzucker, H.J. K^+ Efflux and Retention in Response to NaCl Stress Do Not Predict Salt Tolerance in Contrasting Genotypes of Rice (*Oryza sativa* L.). *PLoS ONE* **2013**, *8*, e57767. [CrossRef]

57. Osakabe, Y.; Yamaguchi-Shinozaki, K.; Shinozaki, K.; Tran, L.P. ABA control of plant macroelement membrane transport systems in response to water deficit and high salinity. *New Phytol.* **2014**, *202*, 35–49. [CrossRef] [PubMed]

58. Martínez-Zamora, L.; Castillejo, N.; Gómez, P.; Artés-Hernández, F. Amelioration Effect of LED Lighting in the Bioactive Compounds Synthesis during Carrot Sprouting. *Agronomy* **2021**, *11*, 304. [CrossRef]

59. Loi, M.; Villani, A.; Paciolla, F.; Mulè, G.; Paciolla, C. Challenges and Opportunities of Light-Emitting Diode (LED) as Key to Modulate Antioxidant Compounds in Plants. A Review. *Antioxidants* **2020**, *10*, 42. [CrossRef] [PubMed]

60. Wang, P.; Chen, S.; Gu, M.; Chen, X.; Chen, X.; Yang, J.; Zhao, F.; Ye, N. Exploration of the Effects of Different Blue LED Light Intensities on Flavonoid and Lipid Metabolism in Tea Plants via Transcriptomics and Metabolomics. *Int. J. Mol. Sci.* **2020**, *21*, 4606. [CrossRef] [PubMed]

61. Bandaranayake, W. Bioactivities, bioactive compounds and chemical constituents of mangrove plants. *Wetl. Ecol. Manag.* **2002**, *10*, 421–452. [CrossRef]

62. AbdElgawad, H.; Zinta, G.; Hegab, M.M.; Pandey, R.; Asard, H.; Abuelsoud, W. High Salinity Induces Different Oxidative Stress and Antioxidant Responses in Maize Seedlings Organs. *Front. Plant Sci.* **2016**, *7*, 276. [CrossRef] [PubMed]

63. Ksouri, R.; Megdiche, W.; Debez, A.; Falleh, H.; Grignon, C.; Abdelly, C. Salinity effects on polyphenol content and antioxidant activities in leaves of the halophyte Cakile maritima. *Plant Physiol. Biochem.* **2007**, *45*, 244–249. [CrossRef] [PubMed]

64. De Abreu, I.N.; Mazzafera, P. Effect of water and temperature stress on the content of active constituents of Hypericum brasiliense Choisy. *Plant Physiol. Biochem.* **2005**, *43*, 241–248. [CrossRef]

65. Certain, C.; Della Patrona, L.; Gunkel-Grillon, P.; Léopold, A.; Soudant, P.; Le Grand, F. Effect of Salinity and Nitrogen Form in Irrigation Water on Growth, Antioxidants and Fatty Acids Profiles in Halophytes *Salsola australis*, *Suaeda maritima*, and *Enchylaena tomentosa* for a Perspective of Biosaline Agriculture. *Agronomy* **2021**, *11*, 449. [CrossRef]

66. Islam, M.Z.; Park, B.-J.; Lee, Y.-T. Effect of salinity stress on bioactive compounds and antioxidant activity of wheat microgreen extract under organic cultivation conditions. *Int. J. Biol. Macromol.* **2019**, *140*, 631–636. [CrossRef]

67. Valifard, M.; Mohsenzadeh, S.; Kholdebarin, B.; Rowshan, V. Effects of salt stress on volatile compounds, total phenolic content and antioxidant activities of Salvia mirzayanii. *South Afr. J. Bot.* **2014**, *93*, 92–97. [CrossRef]

68. Zhou, Y.; Tang, N.; Huang, L.; Zhao, Y.; Tang, X.; Wang, K. Effects of Salt Stress on Plant Growth, Antioxidant Capacity, Glandular Trichome Density, and Volatile Exudates of Schizonepeta tenuifolia Briq. *Int. J. Mol. Sci.* **2018**, *19*, 252. [CrossRef] [PubMed]

69. Ferreyra, M.L.F.; Rius, S.; Emiliani, J.; Pourcel, L.; Feller, A.; Morohashi, K.; Casati, P.; Grotewold, E. Cloning and characterization of a UV-B-inducible maize flavonol synthase. *Plant J.* **2010**, *62*, 77–91. [CrossRef] [PubMed]

70. Dubos, C.; Stracke, R.; Grotewold, E.; Weisshaar, B.; Martin, C.; Lepiniec, L. MYB transcription factors in Arabidopsis. *Trends Plant Sci.* **2010**, *15*, 573–581. [CrossRef]

71. Li, Q.; Kubota, C. Effects of supplemental light quality on growth and phytochemicals of baby leaf lettuce. *Environ. Exp. Bot.* **2009**, *67*, 59–64. [CrossRef]

72. Samuolienė, G.; Sirtautas, R.; Brazaitytė, A.; Sakalauskaitė, J.; Sakalauskienė, S.; Duchovskis, P. The impact of red and blue light-emitting diode illumination on radish physiological indices. *Open Life Sci.* **2011**, *6*, 821–828. [CrossRef]

73. Samuolienė, G.; Viršilė, A.; Brazaitytė, A.; Jankauskienė, J.; Sakalauskienė, S.; Vaštakaitė, V.; Novičkovas, A.; Viškelienė, A.; Sasnauskas, A.; Duchovskis, P. Blue light dosage affects carotenoids and tocopherols in microgreens. *Food Chem.* **2017**, *228*, 50–56. [CrossRef] [PubMed]

74. Zaman, S.; Hu, S.; Alam, A.; Du, H.; Che, S. The accumulation of fatty acids in different organs of purslane under salt stress. *Sci. Hortic.* **2019**, *250*, 236–242. [CrossRef]

75. Munns, R.; Tester, M. Mechanisms of salinity tolerance. *Annu. Rev. Plant Biol.* **2008**, *59*, 651–681. [CrossRef] [PubMed]

76. Ma, R.; Thomas-Hall, S.R.; Chua, E.; Alsenani, F.; Eltanahy, E.; Netzel, M.E.; Netzel, G.; Lu, Y.; Schenk, P.M. Gene expression profiling of astaxanthin and fatty acid pathways in Haematococcus pluvialis in response to different LED lighting conditions. *Bioresour. Technol.* **2018**, *250*, 591–602. [CrossRef]

77. Sharma, N.; Fleurent, G.; Awwad, F.; Cheng, M.; Meddeb-Mouelhi, F.; Budge, S.M.; Germain, H.; Desgagné-Penix, I. Red Light Variation an Effective Alternative to Regulate Biomass and Lipid Profiles in Phaeodactylum tricornutum. *Appl. Sci.* **2020**, *10*, 2531. [CrossRef]

78. He, M.; Ding, N.-Z. Plant Unsaturated Fatty Acids: Multiple Roles in Stress Response. *Front. Plant Sci.* **2020**, *11*, 562785. [CrossRef]
79. Jaffel-Hamza, K.; Sai-Kachout, S.; Harrathi, J.; Lachaâl, M.; Marzouk, B. Growth and Fatty Acid Composition of Borage (*Borago officinalis* L.) Leaves and Seeds Cultivated in Saline Medium. *J. Plant Growth Regul.* **2013**, *32*, 200–207. [CrossRef]
80. Palaniswamy, U.R.; McAvoy, R.J.; Bible, B.B. Stage of harvest and polyunsaturated essential fatty acid concentrations in purslane (*Portulaca oleraceae*) leaves. *J. Agric. Food Chem.* **2001**, *49*, 3490–3493. [CrossRef]

Article

Unraveling the Modulation of Controlled Salinity Stress on Morphometric Traits, Mineral Profile, and Bioactive Metabolome Equilibrium in Hydroponic Basil

Giandomenico Corrado, Paola Vitaglione, Pasquale Chiaiese and Youssef Rouphael *

Department of Agricultural Sciences, University of Naples Federico II, 80055 Portici, Italy; giandomenico.corrado@unina.it (G.C.); paola.vitaglione@unina.it (P.V.); pasquale.chiaiese@unina.it (P.C.)
* Correspondence: youssef.rouphael@unina.it; Tel.: +39-081-2539134

Abstract: Salinity is a major concern in several ecosystems and has a significant impact on global agriculture. To increase the sustainability of horticultural food systems, better management and usage of saline water and soils need to be supported by knowledge of the crop-specific responses to tolerable levels of salinity. The aim of this work was to study the effects of mild salinity on morphological growth and development, leaf color, mineral composition, antioxidant activities, and phenolic profile of sweet basil (*Ocimum basilicum* L.). Plants grew in hydroponics and were exposed to three nutrient solutions (NSs) differing in the NaCl concentration (either 0, 20, or 40 mM). Inhibitory effects on leaf area, fresh yield, and shoot biomass were evident starting from the lowest NaCl concentration, and they became more severe and wide-ranging at 40 mM, also affecting height and root-to-shoot ratio. Salinity increased the nutritional quality in terms of antioxidant activity and polyphenols in leaves, with a reduction in macroelements at 40 mM NaCl. Moreover, the two mild NaCl concentrations specifically modified the concentration of various phenolic acids in leaves. Overall, the use of a slightly saline (20 mM) NS could be tolerated by basil in hydroponics, strongly ameliorating the nutritional profile in the face of relative yield loss. Considering the significantly higher accumulation of bioactive compounds, our work implies that the use of low-salinity water can sustainably increase the nutritional value and the health-promoting features of basil leaves.

Keywords: *Ocimum basilicum*; salt; NaCl; yield; quality; nitrate; minerals; antioxidants; polyphenols

Citation: Corrado, G.; Vitaglione, P.; Chiaiese, P.; Rouphael, Y. Unraveling the Modulation of Controlled Salinity Stress on Morphometric Traits, Mineral Profile, and Bioactive Metabolome Equilibrium in Hydroponic Basil. *Horticulturae* **2021**, *7*, 273. https://doi.org/10.3390/horticulturae7090273

Academic Editor: Lilian Schmidt

Received: 2 August 2021
Accepted: 30 August 2021
Published: 1 September 2021

Publisher's Note: MDPI stays neutral with regard to jurisdictional claims in published maps and institutional affiliations.

Copyright: © 2021 by the authors. Licensee MDPI, Basel, Switzerland. This article is an open access article distributed under the terms and conditions of the Creative Commons Attribution (CC BY) license (https://creativecommons.org/licenses/by/4.0/).

1. Introduction

Salinity is considered a global limiting factor for agriculture, with around one-fifth of irrigated land suffering from secondary salinization [1]. In plants, the salt stress response encompasses multidimensional changes at molecular, cellular, and physiological levels, which typically result in a yield decline and, in the most severe cases, in stunted growth or death according to the type, intensity, and duration of the stress [2,3]. Salinity is a concern for irrigated and nonirrigated lands and affects freshwater quality, soil health, biodiversity, and ultimately society [4], with climate change predicted to exacerbate the problems of salinity in agriculture [5]. It is, therefore, necessary to investigate and possibly adapt horticultural production systems to more saline conditions.

It is long established that salinity and other suboptimal environmental conditions (e.g., temperature, light intensity, nutrients, and water) have a strong impact on the accumulation of secondary compounds in plants [6–8]. The array of diverse molecules probably represents the most important functional feature of medicinal and aromatic plant (MAP) species [9]. MAPs are becoming increasingly valuable in agriculture, due to their growing use in the medical (including phytotherapy and aromatherapy), pharmaceutical, cosmetic, food, and nutraceutical sectors [9,10]. For these reasons, understanding the response to salt in aromatic plants is important not only to ensure economic returns and sustainability but also to optimize crop quality.

Basil (*Ocimum basilicum* L.) is an herbaceous species mainly cultivated for the aromatic leaves, consumed fresh or processed, and employed (along with the other areal parts) for the extraction of the essential oil [11]. Currently, *O. basilicum* is the most commercially important species of its genus and among the most popular and extensively cultivated herbs around the world [11]. Moreover, basil has the potential to become a model species to study flavor compounds [12]. Although cultivated in several countries, basil is susceptible to chilling [13] and performs better in sunny, warm areas and in light soils [11]. The rising interest in MAPs has created novel commercial opportunities for basil growers [14,15]. In the last decades, the professional basil cultivation has progressively shifted towards soilless systems (mainly hydroponics) in controlled and closed environments, for the intrinsic suitability of this species (in terms of harvest, cycle, and size) and to ensure a more constant quality of the produce by limiting environmental variability [16]. Hydroponic systems also allow easier management of the complementary irrigation with saline water compared to the cultivation in soil [17]. In Italy, for instance, more than half of the basil production comes from hydroponics, and it is mainly employed for the industrial preparation of the pesto sauce, thus requiring standardized productions [18]. Salinity due to nonessential ions such as Na is a primary problem for the recirculation of the nutrient solution in closed-loop systems, deterring the use of irrigation water of inadequate quality especially around the Mediterranean basin [19].

The effects of salinity on basil are various. This species, also at the seedling stage [20], is considered moderately resistant to salt [21], although significant intraspecific variations are reported [22,23]. High levels of NaCl (e.g., 100 mM) can induce significant yield loss [24,25]. Moreover, in basil, salinity can influence secondary metabolites, such as polyphenolics [26,27], volatile organic compounds [28], and carotenoids [24], as well as antioxidant activities [23]. This work aimed to have a wide-ranging view of the response of basil to moderate salt stress, under the premise that the adaptation to mild stress in MAPs has critical and possibly contrasting effects on the quantitative and qualitative features of the plant [29]. Considering that the regular consumption of polyphenols is highly appreciated because of their potential health benefits, also in relation to the prevention of some chronic diseases, we investigated morphological traits, leaf elemental composition, and antioxidant activities and polyphenols (by HPLC) in two saline concentrations (20 and 40 mM NaCl) that are well below the toxicity level for basil. Our specific interest was to understand the interplay between biomass reduction and increase in the antioxidant activity and polyphenolic accumulation, to sustainably optimize yield and nutritional quality in hydroponics. This knowledge is relevant to provide indications for better exploitation of basil according to specific needs and markets and to develop more environmentally sound water resource management strategies.

2. Materials and Methods

2.1. Plant Material, Growing Conditions, and Experimental Design

Plants grew in a nonheated glasshouse at "Azienda Agraria e Zootecnica Torre Lama" (Bellizzi, SA, Italy), Department of Agricultural Sciences, University of Naples Federico II, from October to December 2014. Seeds of the basil (*Ocimum basilicum* L.) variety 'Gecom' (S.A.I.S. S.p.a., Cesena, FC, Italy), a Genovese type, were sown in vermiculite. On 27 October, at the three true leaves stage, plants were transplanted to plastic pots filled with approximately 1.3 L of a Brill 3 substrate (Brill Substrate, Georgsdorf, Germany) with a density of 23 plants per square meter. A fresh nutrient solution (NS) was employed at each irrigation treatment with a drip system at a 2 L/h flow rate. We tested three levels of NaCl concentration in a randomized block design. Each experimental unit was made of 15 plants and was replicated three times, for a total of 135 plants. The base NS (0 mM NaCl) was prepared with deionized water and had the following concentrations of mineral nutrients: 13.0 mM $N\text{-}NO_3^-$, 1.0 mM $N\text{-}NH_4^+$, 1.75 mM S, 1.5 mM P, 5.0 mM K, 4.5 mM Ca, 2.0 mM Mg, 20 µM Fe, 9.0 µM Mn, 0.3 µM Cu, 1.6 µM Zn, 20.0 µM B, and 0.3 µM Mo. The two saline treatments were applied using the same base NS supplemented with either

20 mM NaCl or 40 mM NaCl. The pH of the NSs was 6.0. The electrical conductivity (EC) was 2.1 dS/m for the base NS, 3.9 dS/m for the 20 mM NaCl, and 5.8 dS/m for the 40 mM NaCl. Saline treatment started eight days after transplant (DAT) and the growing cycle ended 44 DAT, at the preflowering stage (i.e., commercial maturity).

2.2. Biometric Measurements and Antioxidant Activity

Leaves, stems, and roots from ten plants per block (i.e., replicate) were harvested 39 DAT (4 December) and weighed (fresh weight (fw)) with an analytical balance (Denver Instruments, Denver, CO, USA). Samples were bagged and then dried at 70 °C until constant weight (dry weight (dw)). The total leaf area was measured from ten plants per block using a portable leaf area meter (Li-Cor3000, Li-Cor, Lincoln, NE, USA). The SPAD index was measured on 20 fully expanded leaves per experimental unit with a portable chlorophyll meter (SPAD-502, Konica Minolta, Tokyo, Japan), while color was measured with a CR-300 Chroma Meter (Minolta, Tokyo, Japan) as previously reported [30]. The pairwise difference in color (ΔE^*_{ab}) between treatments was calculated as described [31]. Nitrate content in leaves was evaluated with previously described procedures [30]. The mineral profile (P, K, Ca, Mg, and Na) of the dried leaves was analyzed with an inductively coupled plasma (ICP) emission spectrophotometer (Iris, Thermo Optek, Milan, Italy) as reported [32]. Fresh sweet basil leaves from three plants per experimental unit were immediately frozen in liquid nitrogen after harvest and lyophilized in an Alpha 1–4 LSC plus (Osterode, Germany) freeze drier. The hydrophilic antioxidant activity (HAA) was assessed also as described previously [30].

2.3. Determination of Phenolic Compounds

For the biochemical analyses, we assayed leaves from three plants per experimental block. Phenolics were extracted essentially as described [33]. Briefly, 500 mg of dried leaves was added to 5 mL of methanol:water 30:70 (*v:v*). Samples were thoroughly mixed for 1 min, sonicated for 30 min, centrifuged at 14,800 rpm for 10 min, and then the supernatant was filtered on paper (Whatman, Little Chalfont, UK). The concentrations of caffeoyltartaric acid (CTA), caffeic acid (CA), *p*-coumaric acid (CUA), ferulic acid (FA), chicoric acid (CHA), quercetin rutinoside acid (QRT), and rosmarinic acid (RA) were assessed on the clear supernatant by LC/MS/MS analysis (injection volume: 20 µL). Each sample was extracted and analyzed in triplicate. Chromatographic separation was performed in an HPLC apparatus equipped with two micropumps, Series 200 (Perkin-Elmer Italia, Milan, Italy) and with a Prodigy ODS3 100 Å column (250 mm × 4.6 mm, 5 µm) (Phenomenex, Torrance, CA, USA). The eluents were 0.2% formic acid in water (A) and acetonitrile:methanol 60:40 (*v:v*) (B). The gradient program was 20–30% B (6 min), 30–40% B (10 min), 40–50% B (8 min), 50–90% B (8 min), 90–90% B (3 min), and 90–20% B (3 min) at a constant flow of 0.8 mL/min. MS and MS/MS analyses were carried out on a triple quadrupole mass spectrometer (API 3000, Applied Biosystems Italia, Milan, Italy) equipped with a Turbo Ion Spray electrospray ion source working in the negative ion mode. Compounds were identified by information-dependent acquisition considering the molecular weight, the fragmentation pattern, the retention time, and UV absorption in comparison with those of commercial standards purchased from Sigma-Aldrich (Milan, Italy). The precursor ion and the MS/MS product ions of the phenolic compounds retrieved in the extracts are reported in Table S1. Reagents and solvents (all HPLC grade) were obtained from Merck KGaA (Darmstadt, Germany).

2.4. Statistical Analysis

Measurements of the morphological traits were carried out on ten basil plants per experimental block (made of 15 plants), each replicated three times. The effect of the categorical independent variable "Salt" (three levels: 0, 20 and 40 mM NaCl) on each of the continuous dependant variables was examined using the within-subjects one-way analysis of variance (ANOVA), considering a $p < 0.05$ threshold for statistical significance. Tukey's honestly significant difference (HSD) post hoc test was employed to determine which groups in the ANOVA differ from each other. These calculations were performed with the SPSS 26 software package (IBM, Akron, NY, USA). Data are reported as mean $\pm$ standard error of the mean (s.e.). Principal component analysis (PCA) was carried out on all the measured variables by using XLSTAT (Addinsoft, Paris, France).

3. Results

3.1. Effect on the Plant Morphology

The impact of the mild salinity treatment on the morphological traits of basil was prominent, affecting most of the measured parameters (8 out of 11; Table S2). Moreover, statistical differences between the two NaCl concentrations (i.e., 20 and 40 mM) were not always present. The NaCl salt lowered the plant height at 40 mM (-12%), with a nonsignificant reduction at 20 mM (Figure 1A). Nonetheless, the number of leaves per plant was not affected (Figure 1B), suggesting that plants were stunted rather than less developed. This is also supported by the fact that the two saline treatments strongly restricted the leaf area, which, at the highest NaCl concentration employed, was around one-third of the control condition (Figure 1C). Therefore, the fresh yield was reduced in a similar proportion of the leaf area (around 31% at 40 mM NaCl) (Figure 1D). Interestingly, the effect of the salt on the total shoot biomass was also similar in quantitative terms, with the two saline conditions not significantly different (Figure 1E). The analysis of plant growth in terms of dry biomass indicated the overall inhibitory effect of the NaCl, with a reduction (-19%) from 0 to 20 mM NaCl and a limited (-6%) and nonsignificant difference between 20 and 40 mM NaCl (Figure 1F). This was due to an effect on leaves rather than on stem dry weight (Table S2). The edaphic stress impacted the ratio between roots and shoots, with plants progressively safeguarding the production of root biomass with increasing salt concentration ($+23\%$ and $+46\%$ for 20 and 40 mM NaCl, respectively) (Figure 1G), as indicated by the fact that root biomass was not reduced by NaCl (Table S2). The stress imposed by the salt also affected the percentage of dry mass of the leaves, which significantly increased ($+8\%$) only at 40 mM NaCl (Figure 1H). Overall, the NaCl had an impact on plant height only at the highest saline concentration, and this reduction did not affect the number of leaves. In saline conditions, the yield decrease is due to smaller leaves with a reduced water content, a likely consequence of the inhibitory effect on shoot growth (in terms of dry biomass) and the resulting increase in the root-to-shoot biomass.

3.2. Effect on SPAD Index and Color Coordinates

The salt treatment did not influence the SPAD units, a leaf index related to the chlorophyll content, but did influence the lightness and the chromatic component a* (Table S3). Post hoc analysis indicated that only the 40 mM NaCl treatment significantly differed from the other two experimental conditions. Specifically, at the higher NaCl concentration, leaf lightness moderately rose, and the CIELAB parameter a* increased (e.g., it was less negative), indicating a diminished greenness of the leaves (Figure 2). The overall change in visual perception induced by the salt over the leaf color was evaluated considering the ΔE^*_{ab}. As expected, this value reached the maximum (2.51) comparing the 0 and 40 mM conditions; however, this discrepancy is only slightly above the limit (i.e., 2.0) of a color difference that is perceptible through close observation.

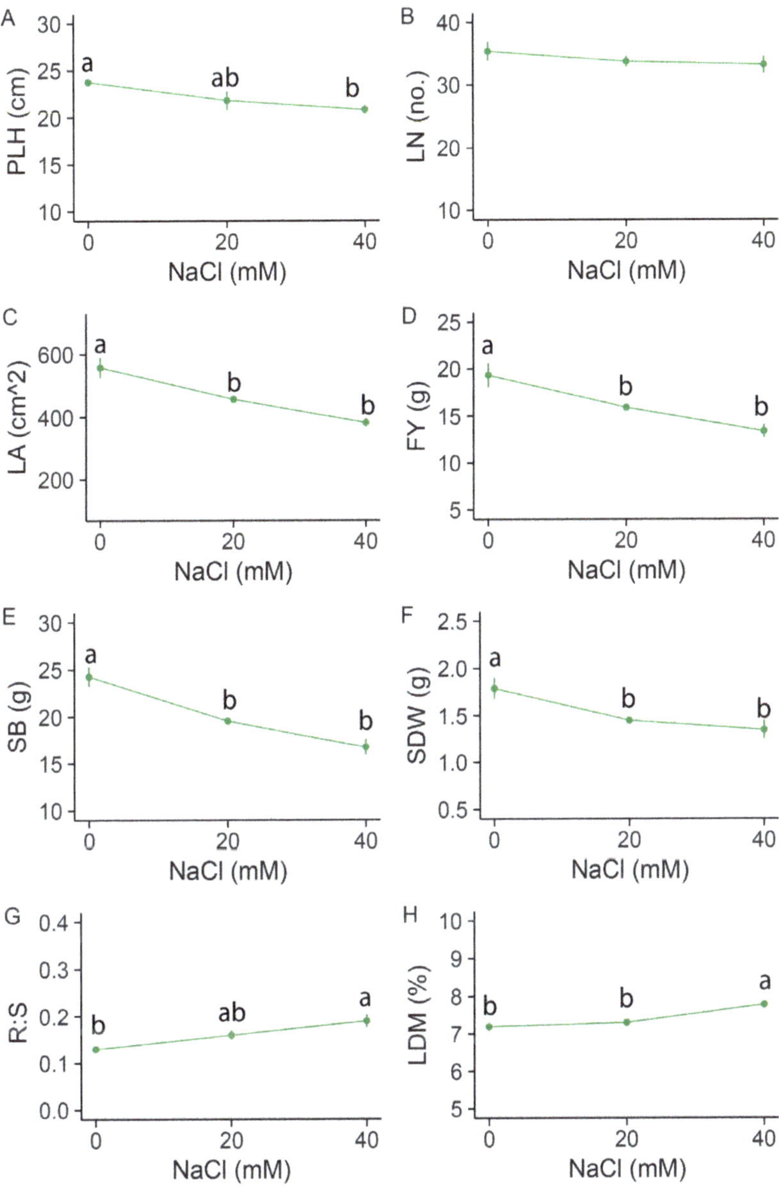

Figure 1. Effects of the NaCl in the nutrient solution on the morphometric traits of basil: (**A**) plant height (PLH); (**B**) leaf number (LN); (**C**) leaf area (LA); (**D**) fresh yield (FY); (**E**) shoot biomass (SB); (**F**) shoot dry weight (SDW); (**G**) root-to-shoot ratio (R:S); (**H**) leaf dry matter (LDM). Data are mean ± standard error of the mean. Different letters indicate significantly different means according to the Tukey HSD test (*p* < 0.05).

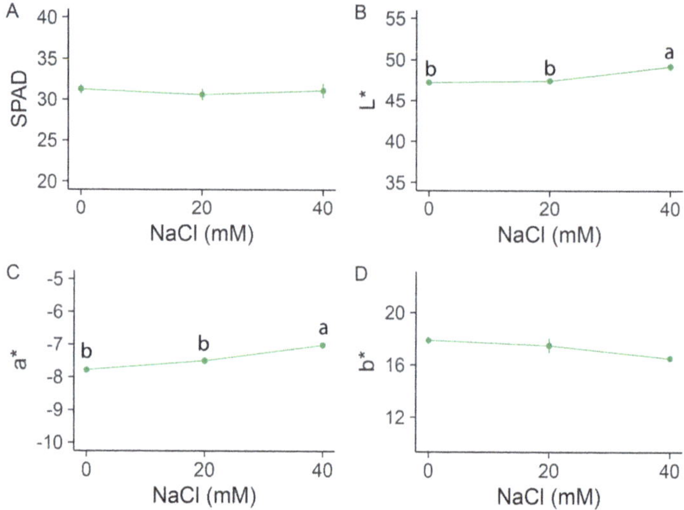

Figure 2. Effects of the NaCl in the nutrient solution on SPAD index and the L*, a*, and b* color coordinates according to the CIELAB standard: (**A**) SPAD; (**B**) L*; (**C**) a*; (**D**) b*. Data are mean ± standard error of the mean. Different letters indicate significantly different means according to the Tukey HSD test ($p < 0.05$).

3.3. Effects on the Elemental Composition of the Leaves

We measured the nitrate content in basil leaves because this parameter influences the nutritional quality of this species and, more generally, of leafy greens. Both the 20 and 40 mM NaCl treatments significantly reduced the nitrates in leaves (Table S4), on average, by 23% compared to the control conditions (Figure 3A). NaCl also affected the amount of all the mineral elements analyzed (Table S4). Specifically, the strongest reduction was observed for Ca (−38% and −54% for the 20 and 40 mM NaCl, respectively) (Figure 3D). For P, K, and Mg, only the 40 mM dose caused a statistically significant reduction compared to the control condition, although a declining trend according to increasing NaCl concentration could be inferred (Figure 3B,C,E). K was the element that little changed relative to the control condition (−16% and −32% for the 20 and 40 mM NaCl, respectively). As expected, the amount of Na in leaves rose upon application of increasingly saline NS (Figure 3F). Na was the only analyzed element that accumulated in statistically different amounts moving from the 20 mM to the 40 mM NaCl treatment.

3.4. Effects on HAA and Total Polyphenols

Considering the importance of the antioxidants for leafy greens and, specifically, the phytotherapeutic use of basil, we measured the antioxidant activity of the aqueous extract, and total polyphenols because they are the most abundant phytochemicals in the human diet that can remove oxidizing agents in cells [34]. Salinity affected both parameters (Table S5). It was noteworthy that NaCl increased both traits, on average +77% for the HAA and +201% for the polyphenols (Figure 4A,B, respectively). Moreover, this increase in the nutritional quality of the basil leaf was not different when the two saline conditions were compared.

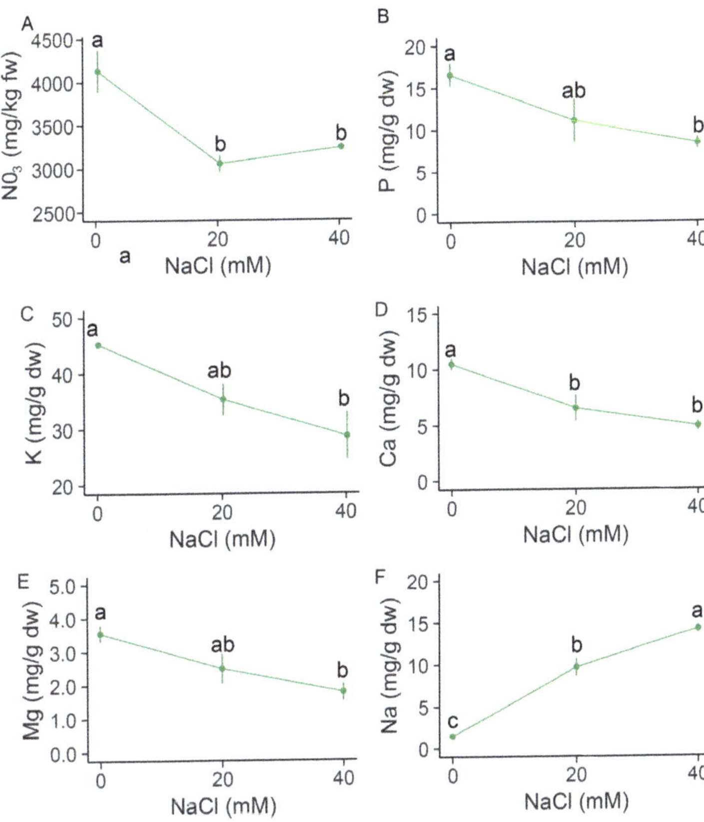

Figure 3. Effects of the NaCl in the nutrient solution on the elemental composition of the basil leaves: (**A**) nitrate; (**B**) P; (**C**) K; (**D**) Ca; (**E**) Mg; (**F**) Na. Data are mean ± standard error of the mean. Different letters indicate significantly different means according to the Tukey HSD test ($p < 0.05$).

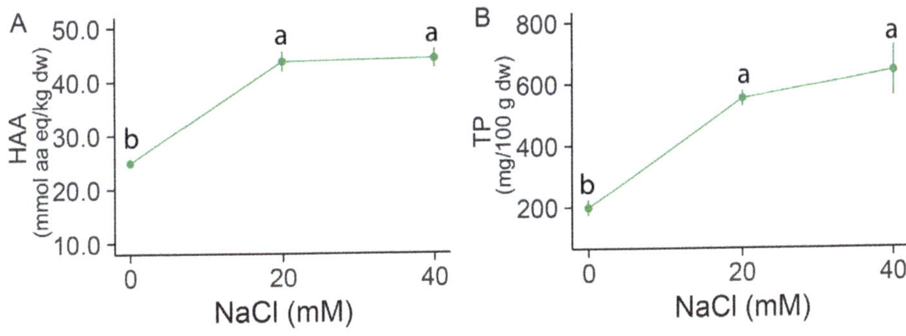

Figure 4. Effects of the NaCl in the nutrient solution on the hydrophilic antioxidant activity (HAA) and the total polyphenols (TP) in leaves: (**A**) HAA; (**B**) TP. Data are mean ± standard error of the mean. Different letters indicate significantly different means according to the Tukey HSD test ($p < 0.05$).

3.5. Effect on the Polyphenolic Profile of Basil Leaves

Considering the known relation between plant stress-response and the accumulation of secondary compounds, we tested if, also under mild saline conditions, different NaCl concentrations specifically alter the accumulation of polyphenols. The saline treatment varied the accumulation of five out of the seven detected compounds, irrespective of the absolute amount of the various polyphenols (Table S5). It was interesting that different and, in some cases, divergent responses were recorded, despite the overall positive effect of the salt stress on their total amount. For instance, the quercetin rutinoside acid was the only polyphenol that constantly increased with higher saline conditions, reaching a 2× higher concentration at 40 mM NaCl (Figure 5F). At this experimental condition, the concentrations of *p*-coumaric acid, chicoric acid, and rosmarinic acid (the major caffeic acid ester) were also significantly higher than the control condition (Figure 5B,D,G). Among them, there was a statistical difference between the two saline conditions only for the *p*-coumaric acid. The highest relative increase was observed for chicoric acid (just above 10× higher at 40 mM than the control condition). Ferulic acid was the only compound that peaked at 20 mM (5× higher than in the control leaves), while its concentration at 40 mM did not differ from the no salt treatment (Figure 5E). At 20 mM, ferulic acid became the predominant phenolic acid in basil leaves. Overall, this analysis indicates that while any applied salt stress increased the sum of polyphenols, the two different mild NaCl concentrations employed specifically modified the phenolic acid composition in basil leaves.

To summarize the information content of the different analyses, we performed a principal component analysis (PCA) on all the measured variables. The first two principal components were held to graphically visualize the relation between the experimental conditions. The first two principal components were associated with eigenvalues higher than 1 and explained 100% of the cumulative variance, with the first and second principal components accounting for 81.7% and 18.3%, respectively. The multivariate analysis indicated that the different treatments are well separated, with the 0 mM NaCl condition more distant from the other two salt concentrations (Figure 6). The control treatment (0 mM NaCl) was distinguished for morphometric traits (fresh yield, shoot biomass, shoot dry weight, leaf area and plant height) and for cations (K, Ca and Mg) and nitrate content (Figure 6). Interestingly, sweet basil treated with 20 or 40 mM NaCl delivered leaves with high concentration of total polyphenols and hydrophilic antioxidant activity (Figure 6). Finally, sweet basil irrigated with 40 mM NaCl was positioned on the negative side of the first principal component in the lower left quadrant of the PCA score plot, characterized overall by higher target polyphenols (chicoric acid, *p*-coumaric acid, quercetin rutinoside acid) and color coordinates (L* and a*).

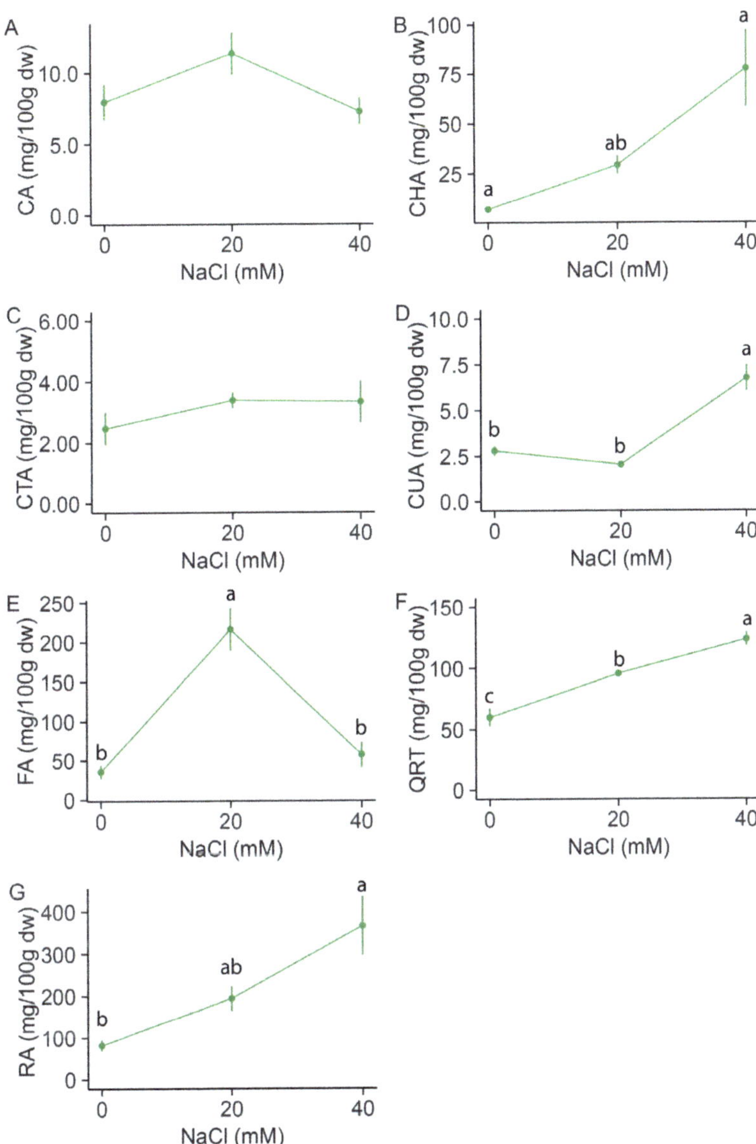

Figure 5. Effects of the NaCl in the nutrient solution on the polyphenolic compounds in leaves: (**A**) caffeic acid (CA); (**B**) chicoric acid (CHA); (**C**) caffeoyltartaric acid (CTA); (**D**) *p*-coumaric acid (CUA); (**E**) ferulic acid (FA); (**F**) quercetin rutinoside acid (QRT); (**G**) rosmarinic acid (RA). Data are mean $\pm$ standard error of the mean. Different letters indicate significantly different means according to the Tukey HSD test ($p < 0.05$).

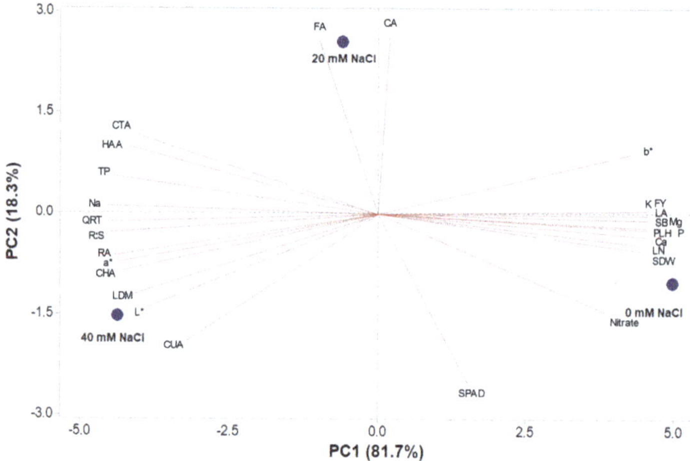

Figure 6. Principal component analysis (PCA) of the basil response to the three experimental conditions, 0, 20, and 40 mM NaCl. Plant height (PLH), leaf number (LN), leaf area (LA), fresh yield (FY), shoot biomass (SB), shoot dry weight (SDW), root-to-shoot ratio (R:S), leaf dry matter (LDM), hydrophilic antioxidant activity (HAA), total polyphenols (TP), caffeic acid (CA), chicoric acid (CHA), caffeoyltartaric acid (CTA), *p*-coumaric acid (CUA), ferulic acid (FA), quercetin rutinoside acid (QRT), rosmarinic acid (RA).

4. Discussion

In addition to its historical position in the national gastronomy, the professional cultivation of basil in Italy is economically sustained by the food industry, with large-scale production occurring mainly in hydroponics. Soilless systems offer the possibility to tailor the quality of the edible fraction by modulating preharvest factors, including the electric conductivity of the nutrient solution. This is relevant also because, in recent years, the growing concerns about the availability and quality of freshwater have expanded the need for a more comprehensive evaluation of the effects of a mild NaCl stress on the yield and quality of basil, as well as other MAP species [35,36].

The vegetative growth of the basil plants was negatively affected by the salinity, while development (measured by the number of leaves per plant) was not influenced. The saline treatment decreased fresh yield because of smaller leaves with reduced water content. Salinity is a known inhibitor of plant growth, as also reported in other aromatic Lamiaceae such as sage [37], spearmint [38], and marjoram [39]. Specifically, the reduction in the leaf area is one of the first and most recorded symptoms of the salinity in stress in plants [40], and our data added that in basil, this parameter is the most highly responsive in mild NaCl concentrations. In basil and other Lamiaceae, the effects of salt stress can be more severe and also include a significantly lower number of leaves, in addition to the reduction in the leaf area and leaf biomass [24,41]. Nonetheless, in basil the number of internodes was little reduced only from 75 mM NaCl [24]. On the other hand, in peppermint, the leaf fresh weight and area were reduced at EC values of 2.8 and 5.6 dS/m, respectively [41]. The inhibitory effect of salt on shoot dry biomass along with the limited effect on roots resulted in elevated root-to-shoot biomass. In a previous study, only a concentration of seawater higher than 40% limited root growth in basil, while an increased root-to-shoot ratio was observed starting from 5% [42]. Finally, salinity had a small effect on leaf color, confirming that the saline conditions were not close to a toxicity level [27].

The inhibitory effect of the saline treatments cannot be disconnected from the alteration of nutrient uptake and transport. Our data indicated that the elemental composition of the leaves was highly affected. All the analyzed mineral nutrients (present in a fixed amount in the NSs) were reduced at increasing NaCl concentrations, consistent with the higher uptake and translocation in leaves of the NaCl. The antagonist ionic relation between Na^+ and Ca^{2+} or K^+ and between Cl^- and NO_3^- is a key factor affecting the mineral composition of the plants under NaCl stress [43–45]. This effect was evident also in mild concentrations, as also occurred in *Mentha pulegium*, where a significant inhibition on K^+ was evident starting from 25 mM NaCl [46]. It was noteworthy that nitrate reduction did not well correlate with Cl concentration in leaves, suggesting the mechanisms other than anion–anion antagonism may account for the observed phenomenon [44,47]. The significant reduction in nitrate at the lowest NaCl concentration also has dietary implications because a limited accumulation of this anion in edible products is desired to avoid potential health hazards.

The NaCl also increased the quality of basil in terms of hydrophilic antioxidant activity and polyphenols, which suggests that the observed effects are due also to oxidative stress, and not only to ion antagonisms and osmotic imbalance [3]. Salt stress disturbs cellular homeostasis in different ways, with consequent oxidative damage associated with an overproduction of reactive oxygen species (ROS) [3]. In our mild salinity concentrations, basil plants can protect themselves by increasing the antioxidant activity and polyphenols, a major biochemical class of antioxidants in plants. From an applied perspective, the data indicated that a positive effect on the quality attributes of the basil leaves can be obtained with the lowest NaCl concentration (20 mM), and no further benefits were obtained by moving to the most severe suboptimal conditions. In rosemary, both antioxidant activities and total phenols at 50 and 100 mM NaCl were higher than the control condition, and they both decreased at 150 mM NaCl [48].

While several reports indicated that plants generally accumulate phenolics during salt stress [49], relatively little is known about the effect on specific compounds. While our analysis confirmed that rosmarinic acid is a predominant phenolic compound in "Genovese" basil [50], the data also indicated that the large quantitative variation due to NaCl is accompanied by different accumulations of specific compounds. For instance, two hydroxycinnamic acid derivatives, caffeic acid and caffeoyltartaric acid, were not affected by the NaCl treatment, but *p*-coumaric acid, present in a similarly low amount, strongly increased at 40 mM. Moreover, a substantial dose-dependent variation was present for two major components, ferulic acid and rosmarinic acid. While the latter showed a constant increase with higher salt concentration, the former peaked at 20 mM. Overall, the impact on salinity on the polyphenol composition appears to be more complex and larger than on other measured parameters, although NaCl treatments never caused a decrease in specific polyphenols. The relative increase in the concentration of major polyphenols in leaves was above the biomass reduction, suggesting an increasing biosynthesis and accumulation of these compounds. The polyphenol accumulation in basil may also vary according to the cultivar or type [27,50], and further studies may verify the importance of the genetic factor in the response of individual compounds to NaCl.

5. Conclusions

The multidimensional evaluation of the response of the plants under mild salinity conditions is a necessary step to increase the environmental sustainability and quality of basil. This work confirmed the complex impact of salinity on hydroponic basil and provided some indications on the factors that should be considered when employing mildly saline irrigation water. Basil can be considered moderately salt-tolerant, with a little growth reduction occurring at 20 mM NaCl and a reduced yield mainly due to smaller leaves. This saline concentration offered the most interesting results based on the magnitude of the morphological alterations and the increase in the leaf quality in terms of nitrate content, antioxidant activity, and polyphenol profile. Considering the expected positive impact on the nutraceutical properties of basil and, potentially, on its market value, our study encourages more detailed analyses of the health-related properties of basil following the implementation of nonconventional (saline or sodic) water sources in cultivation.

Supplementary Materials: The following are available online at https://www.mdpi.com/article/10.3390/horticulturae7090273/s1, Table S1: Table S1. LC/MS/MS characteristics of phenolic compounds identified and quantified in the basil extracts. Table S2: One-way ANOVA table of the morphometric traits. The unit of measurement of each variable is reported in brackets. Table S3: One-way ANOVA table of the SPAD index and color parameters. Table S4: One-way ANOVA table of the mineral content in basil leaves. The unit of measurement of each variable is reported in brackets. Table S5: One-way ANOVA table of the biochemical traits investigated. The unit of measurement of each variable is reported in brackets.

Author Contributions: Conceptualization, G.C. and Y.R.; methodology, P.V. and Y.R.; software, G.C. and Y.R.; formal analysis, G.C., P.V., P.C., and Y.R.; investigation, G.C. and Y.R.; resources, Y.R.; data curation G.C. and Y.R.; writing—original draft preparation, G.C.; writing—review and editing, G.C., P.V., P.C., and Y.R.; visualization, G.C., P.V., P.C., and Y.R.; supervision, G.C. and Y.R.; project administration Y.R. All authors have read and agreed to the published version of the manuscript.

Funding: This research received no external funding.

Institutional Review Board Statement: Not applicable.

Informed Consent Statement: Not applicable.

Data Availability Statement: The data presented in this study are available in the article and the Supplementary Material. Raw data are available on request from the corresponding author.

Acknowledgments: The authors are grateful to Emilio Di Stasio and Giampaolo Raimondi for their assistance in the greenhouse experiment.

Conflicts of Interest: The authors declare no conflict of interest.

References

1. Flowers, T.; Yeo, A. Breeding for Salinity Resistance in Crop Plants: Where Next? *Funct. Plant Biol.* **1995**, *22*, 875–884. [CrossRef]
2. Van Zelm, E.; Zhang, Y.; Testerink, C. Salt Tolerance Mechanisms of Plants. *Annu. Rev. Plant Biol.* **2020**, *71*, 403–433. [CrossRef] [PubMed]
3. Munns, R.; Tester, M. Mechanisms of salinity tolerance. *Annu. Rev. Plant Biol.* **2008**, *59*, 651–681. [CrossRef]
4. Pitman, M.G.; Läuchli, A. Global Impact of Salinity and Agricultural Ecosystems. In *Salinity: Environment-Plants-Molecules*; Springer: Berlin/Heidelberg, Germany, 2002; pp. 3–20.
5. Corwin, D.L. Climate change impacts on soil salinity in agricultural areas. *Eur. J. Soil Sci.* **2021**, *72*, 842–862. [CrossRef]
6. Miras-Moreno, B.; Corrado, G.; Zhang, L.; Senizza, B.; Righetti, L.; Bruni, R.; El-Nakhel, C.; Sifola, M.I.; Pannico, A.; De Pascale, S.; et al. The Metabolic Reprogramming Induced by Sub-optimal Nutritional and Light Inputs in Soilless Cultivated Green and Red Butterhead Lettuce. *Int. J. Mol. Sci.* **2020**, *21*, 6381. [CrossRef] [PubMed]
7. Ramakrishna, A.; Ravishankar, G. Role of Plant Metabolites in Abiotic Stress Tolerance Under Changing Climatic Conditions with Special Reference to Secondary Compounds. In *Climate Change and Plant Abiotic Stress Tolerance*; Wiley: Hoboken, NJ, USA, 2013; pp. 705–726. [CrossRef]
8. Obata, T.; Fernie, A.R. The use of metabolomics to dissect plant responses to abiotic stresses. *Cell. Mol. Life Sci.* **2012**, *69*, 3225–3243. [CrossRef]
9. Bajaj, Y.P.S. *Medicinal and Aromatic Plants I*; Springer Science & Business Media: Berlin/Heidelberg, Germany, 2012; Volume 4.
10. Padulosi, S.; Leaman, D.; Quek, P. Challenges and Opportunities in Enhancing the Conservation and Use of Medicinal and Aromatic Plants. *J. Herbs Spices Med. Plants* **2002**, *9*, 243–267. [CrossRef]
11. Makri, O.; Kintzios, S. Ocimumsp. (Basil): Botany, Cultivation, Pharmaceutical Properties, and Biotechnology. *J. Herbs, Spices Med. Plants* **2008**, *13*, 123–150. [CrossRef]
12. Gang, D.R. Evolution of flavors and scents. *Annu. Rev. Plant Biol.* **2005**, *56*, 301–325. [CrossRef] [PubMed]
13. Shafran, E.; Dudai, N.; Mayer, A.M. Polyphenol oxidase in *Ocimum basilicum* during growth, development and following cold stress. *J. Food Agric. Environ.* **2007**, *5*, 254.
14. Gunjan, M.; Naing, T.W.; Saini, R.S.; Ahmad, A.; Naidu, J.R.; Kumar, I. Marketing trends & future prospects of herbal medicine in the treatment of various disease. *World J. Pharm. Res.* **2015**, *4*, 132–155.
15. Martin, C.; Zhang, Y.; Tonelli, C.; Petroni, K. Plants, Diet, and Health. *Annu. Rev. Plant Biol.* **2013**, *64*, 19–46. [CrossRef] [PubMed]
16. Tomasi, N.; Pinton, R.; Dalla Costa, L.; Cortella, G.; Terzano, R.; Mimmo, T.; Scampicchio, M.; Cesco, S. New 'solutions' for floating cultivation system of ready-to-eat salad: A review. *Trends Food Sci. Technol.* **2015**, *46*, 267–276. [CrossRef]
17. Atzori, G.; Mancuso, S.; Masi, E. Seawater potential use in soilless culture: A review. *Sci. Hortic.* **2019**, *249*, 199–207. [CrossRef]
18. Tesi, R. *Orticoltura Mediterranea Sostenibile*; Pàtron Editore: Granarolo dell'Emilia, Italy, 2010; pp. 1–503.
19. Neocleous, D.; Ntatsi, G.; Savvas, D. Physiological, nutritional and growth responses of melon (Cucumis meloL.) to a gradual salinity built-up in recirculating nutrient solution. *J. Plant Nutr.* **2017**, *40*, 2168–2180. [CrossRef]

20. Ramin, A.A. Effects of Salinity and Temperature on Germination and Seedling Establishment of Sweet Basil (*Ocimum basilicum* L.). *J. Herbs, Spices Med. Plants* **2006**, *11*, 81–90. [CrossRef]

21. Prasad, A.; Lal, R.K.; Chattopadhyay, A.; Yadav, V.K.; Yadav, A. Response of Basil Species to Soil Sodicity Stress. *Commun. Soil Sci. Plant Anal.* **2007**, *38*, 2705–2715. [CrossRef]

22. Barbieri, G.; Vallone, S.; Orsini, F.; Paradiso, R.; De Pascale, S.; Negre-Zakharov, F.; Maggio, A. Stomatal density and metabolic determinants mediate salt stress adaptation and water use efficiency in basil (*Ocimum basilicum* L.). *J. Plant Physiol.* **2012**, *169*, 1737–1746. [CrossRef] [PubMed]

23. Tarchoune, I.; Sgherri, C.; Baâtour, O.; Izzo, R.; Lachaâl, M.; Navari-Izzo, F.; Ouerghi, Z. Effects of oxidative stress caused by NaCl or Na2SO4 excess on lipoic acid and tocopherols in Genovese and Fine basil (*Ocimum basilicum*). *Ann. Appl. Biol.* **2013**, *163*, 23–32. [CrossRef]

24. Bernstein, N.; Kravchik, M.; Dudai, N. Salinity-induced changes in essential oil, pigments and salts accumulation in sweet basil (*Ocimum basilicum*) in relation to alterations of morphological development. *Ann. Appl. Biol.* **2010**, *156*, 167–177. [CrossRef]

25. Heidari, M. Effects of salinity stress on growth, chlorophyll content and osmotic components of two basil (*Ocimum basilicum* L.) genotypes. *Afr. J. Biotechnol.* **2012**, *11*, 379–384. [CrossRef]

26. Bekhradi, F.; Delshad, M.; Marín, A.; Luna, M.C.; Garrido, Y.; Kashi, A.; Babalar, M.; Gil, M.I. Effects of salt stress on physiological and postharvest quality characteristics of different Iranian genotypes of basil. *Hortic. Environ. Biotechnol.* **2015**, *56*, 777–785. [CrossRef]

27. Scagel, C.F.; Lee, J.; Mitchell, J.N. Salinity from NaCl changes the nutrient and polyphenolic composition of basil leaves. *Ind. Crop. Prod.* **2019**, *127*, 119–128. [CrossRef]

28. Tarchoune, I.; Baâtour, O.; Harrathi, J.; Cioni, P.L.; Lachaâl, M.; Flamini, G.; Ouerghi, Z. Essential oil and volatile emissions of basil (*Ocimum basilicum*) leaves exposed to NaCl or Na$_2$SO$_4$ salinity. *J. Plant Nutr. Soil Sci.* **2013**, *176*, 748–755. [CrossRef]

29. Banerjee, A.; Roychoudhury, A. Effect of Salinity Stress on Growth and Physiology of Medicinal Plants. In *Medicinal Plants and Environmental Challenges*; Springer: Berlin/Heidelberg, Germany, 2017; pp. 177–188.

30. Corrado, G.; Formisano, L.; De Micco, V.; Pannico, A.; Giordano, M.; El-Nakhel, C.; Chiaiese, P.; Sacchi, R.; Rouphael, Y. Understanding the Morpho-Anatomical, Physiological, and Functional Response of Sweet Basil to Isosmotic Nitrate to Chloride Ratios. *Biology* **2020**, *9*, 158. [CrossRef] [PubMed]

31. Lisanti, M.; Mataffo, A.; Scognamiglio, P.; Teobaldelli, M.; Iovane, M.; Piombino, P.; Rouphael, Y.; Kyriacou, M.; Corrado, G.; Basile, B. 1-Methylcyclopropene Improves Postharvest Performances and Sensorial Attributes of Annurca-Type Apples Exposed to the Traditional Reddening in Open-Field *Melaio*. *Agronomy* **2021**, *11*, 1056. [CrossRef]

32. Volpe, M.G.; Nazzaro, M.; Di Stasio, M.; Siano, F.; Coppola, R.; De Marco, A. Content of micronutrients, mineral and trace elements in some Mediterranean spontaneous edible herbs. *Chem. Central J.* **2015**, *9*, 1–9. [CrossRef] [PubMed]

33. Ferracane, R.; Graziani, G.; Gallo, M.; Fogliano, V.; Ritieni, A. Metabolic profile of the bioactive compounds of burdock (Arctium lappa) seeds, roots and leaves. *J. Pharm. Biomed. Anal.* **2010**, *51*, 399–404. [CrossRef]

34. Pandey, K.B.; Rizvi, S.I. Plant Polyphenols as Dietary Antioxidants in Human Health and Disease. *Oxidative Med. Cell. Longev.* **2009**, *2*, 270–278. [CrossRef]

35. Qadir, M.; Boers, T.; Schubert, S.; Ghafoor, A.; Murtaza, G. Agricultural water management in water-starved countries: Challenges and opportunities. *Agric. Water Manag.* **2003**, *62*, 165–185. [CrossRef]

36. Dagar, J.; Minhas, P.; Kumar, M. Cultivation of medicinal and aromatic plants in saline environments. *Plant Sci. Rev.* **2011**, *2012*, 21. [CrossRef]

37. Hendawy, S.; Khalid, K.A. Response of sage (*Salvia officinalis* L.) plants to zinc application under different salinity levels. *J. Appl. Sci. Res.* **2005**, *1*, 147–155.

38. Chrysargyris, A.; Loupasaki, S.; Petropoulos, S.A.; Tzortzakis, N. Salinity and cation foliar application: Implications on essential oil yield and composition of hydroponically grown spearmint plants. *Sci. Hortic.* **2019**, *256*, 108581. [CrossRef]

39. El-Keltawi, N.E.; Croteau, R. Salinity depression of growth and essential oil formation in spearmint and marjoram and its reversal by foliar applied cytokinin. *Phytochemistry* **1987**, *26*, 1333–1334. [CrossRef]

40. Parida, A.K.; Das, A.B. Salt tolerance and salinity effects on plants: A review. *Ecotoxicol. Environ. Saf.* **2005**, *60*, 324–349. [CrossRef] [PubMed]

41. Tabatabaie, S.J.; Nazari, J. Influence of nutrient concentrations and NaCl salinity on the growth, photosynthesis, and essential oil content of peppermint and lemon verbena. *Turk. J. Agric. For.* **2007**, *31*, 245–253.

42. Ning, J.; Cui, L.; Yang, S.; Ai, S.; Li, M.; Sun, L.; Chen, Y.; Wang, R.; Zeng, Z. Basil ionic responses to seawater stress and the identification of gland salt secretion. *J. Anim. Plant Sci.* **2015**, *25*, 131–138.

43. Abdelgadir, E.M.; Oka, M.; Fujiyama, H. Characteristics of Nitrate Uptake by Plants Under Salinity. *J. Plant Nutr.* **2005**, *28*, 33–46. [CrossRef]

44. Corrado, G.; Lucini, L.; Miras-Moreno, B.; Chiaiese, P.; Colla, G.; De Pascale, S.; Rouphael, Y. Metabolic Insights into the Anion-Anion Antagonism in Sweet Basil: Effects of Different Nitrate/Chloride Ratios in the Nutrient Solution. *Int. J. Mol. Sci.* **2020**, *21*, 2482. [CrossRef]

45. Rubinigg, M.; Posthumus, F.; Ferschke, M.; Elzenga, J.M.; Stulen, I. Effects of NaCl salinity on 15N-nitrate fluxes and specific root length in the halophyte *Plantago maritima* L. *Plant Soil* **2003**, *250*, 201–213. [CrossRef]

46. Oueslati, S.; Karray-Bouraoui, N.; Attia, H.; Rabhi, M.; Ksouri, R.; Lachaâl, M. Physiological and antioxidant responses of Mentha pulegium (Pennyroyal) to salt stress. *Acta Physiol. Plant.* **2009**, *32*, 289–296. [CrossRef]
47. Sreenivasulu, N.; Grimm, B.; Wobus, U.; Weschke, W. Differential response of antioxidant compounds to salinity stress in salt-tolerant and salt-sensitive seedlings of foxtail millet (*Setaria italica*). *Physiol. Plant.* **2000**, *109*, 435–442. [CrossRef]
48. Kh, K.; Mohseni, R.; Saboora, A. Biochemical changes of *Rosmarinus officinalis* under salt stress. *J. Stress Physiol. Biochem.* **2010**, *6*, 114–122.
49. Waśkiewicz, A.; Muzolf-Panek, M.; Goliński, P. Phenolic Content Changes in Plants Under Salt Stress. In *Ecophysiology and Responses of Plants under Salt Stress*; Springer: New York, NY, USA, 2013; pp. 283–314.
50. Tarchoune, I.; Sgherri, C.; Izzo, R.; Navari-Izzo, F.; Zeineb, O. Phenolic acids and total antioxidant activity in *Ocimum basilicum* L. grown under Na_2SO_4 medium. *J. Med. Plants Res.* **2012**, *6*, 5868–5875.

horticulturae

MDPI

Article

Effects of Drought on Yield and Nutraceutical Properties of Beans (*Phaseolus spp.*) Traditionally Cultivated in Veneto, Italy

Pietro Sica [1,2], Aline Galvao [2], Francesco Scariolo [2], Carmelo Maucieri [2], Carlo Nicoletto [2], Cristiane Pilon [3], Paolo Sambo [2], Gianni Barcaccia [2], Maurizio Borin [2], Miguel Cabrera [1] and Dorcas Franklin [1,*]

[1] Department of Crop & Soil Sciences, University of Georgia, 3111 Miller Plant Sciences Bldg., Athens, GA 30602, USA; pietros0394@gmail.com (P.S.); mcabrera@uga.edu (M.C.)

[2] Department of Agronomy, Food, Natural Resources, Animals, and Environment (DAFNAE), University of Padova, 16 Viale dell'Università, 35020 Legnaro, Italy; aline.galvao@unipd.it (A.G.); francesco.scariolo@studenti.unipd.it (F.S.); carmelo.maucieri@unipd.it (C.M.); carlo.nicoletto@unipd.it (C.N.); paolo.sambo@unipd.it (P.S.); gianni.barcaccia@unipd.it (G.B.); maurizio.borin@unipd.it (M.B.)

[3] Department of Crop & Soil Sciences, University of Georgia, Rainwater Rd., Tifton, GA 31793, USA; cpilon@uga.edu

* Correspondence: dfrankln@uga.edu; Tel.: +1-706-340-4870

Abstract: Beans are often grown in regions with climates that are susceptible to drought during the cultivation period. Consequently, it is important to identify bean accessions tolerant to drought conditions and assess the effect of drought on seeds' nutraceutical properties. This study evaluated the effect of drought during different development stages (NES = never stressed; ALS = always stressed; SBF = stressed before flowering; SAF = stressed after flowering) on the yield and nutraceutical properties of six local bean varieties: Fasolo del Diavolo, Gialet, Posenati, Secle, D'oro, and Maron. Analysis of variance indicated that Gialet was not significantly affected by drought treatments, and Posenati under SBF and NES treatments had greater yields than under ALS and SAF treatments, whereas Secle under SBF produced 80% more seeds than under NES. Total phenols, antioxidant capacity, and calcium content were significantly different among the local varieties. Yield was significantly and positively correlated with seed calcium content and significantly and negatively correlated with protein, total phenols, and antioxidant capacity. The interaction between local varieties and treatment significantly affected seeds' Zn content. Gialet and Maron seeds' Zn contents were about 60 mg kg^{-1}, almost double the average of commercial varieties. In summary, this study paves the way to the identification of potential bean varieties resistant to drought. Further molecular studies will help support these findings.

Keywords: heirloom beans; drought; abiotic stress; local farming; nutraceutical properties; zinc

Citation: Sica, P.; Galvao, A.; Scariolo, F.; Maucieri, C.; Nicoletto, C.; Pilon, C.; Sambo, P.; Barcaccia, G.; Borin, M.; Cabrera, M.; et al. Effects of Drought on Yield and Nutraceutical Properties of Beans (*Phaseolus spp.*) Traditionally Cultivated in Veneto, Italy. *Horticulturae* **2021**, *7*, 17. https://doi.org/10.3390/horticulturae7020017

Academic Editor: Douglas D. Archbold

Received: 2 December 2020

Accepted: 15 January 2021

Published: 26 January 2021

Publisher's Note: MDPI stays neutral with regard to jurisdictional claims in published maps and institutional affiliations.

Copyright: © 2021 by the authors. Licensee MDPI, Basel, Switzerland. This article is an open access article distributed under the terms and conditions of the Creative Commons Attribution (CC BY) license (https://creativecommons.org/licenses/by/4.0/).

1. Introduction

Legumes are essential to improve global food safety and security [1,2]. Common bean (*Phaseolus vulgaris*) is the most important directly consumed legume in the human diet [3] and the principal food legume for 250 million people in South and Central America and for 400 million people in Eastern Africa [4]. More than 40,000 varieties of beans are cultivated in all continents, except Antarctica, under different cropping systems and in a wide range of environments. Beans are considered one of the major crops with the highest levels of variation in growth habit, seed characteristics, maturity, and adaptation [5]. Beans also play an important role in achieving a more sustainable food production system [6–8] as they are considered an essential crop to help increase food production by 70% by 2050 in order to supply enough high-quality food to the world's growing population sustainably [9].

Legumes and beans are also gaining importance because of their health benefits preventing and helping to manage hypercholesterolemia, hypertension [10], obesity, diabetes, and coronary conditions [6]. Dry bean is known as a major source of protein, with a protein content from 20 to 30% of dry weight [11]. Thus, half a cup of beans (equivalent to 90 g) provides from 18 to 27 g of protein, which is from 32 to 48% of the recommended daily allowances (RDA) of a 70-kg adult (0.8 g of protein per kg of body weight; thus, 56 g) [12,13]. Beans also have considerable contents of phenols and flavonoids. Polyphenols usually are present in the seed coat, rarely in the cotyledons. [14,15]. Colored varieties tend to have a higher content of total phenols and greater antioxidant activities compared to non-pigmented ones [15,16]. Phenols can bring benefits such as reducing the risk of cancer, aging, apoptosis, and oxidative stress [17–19]. Antioxidants can be found in vegetable and fruits and prevent or delay some types of cell damage [20]. Dry bean is considered to be in both the vegetable group and the protein food group, as dry beans have similar protein profiles as foods in both groups and also provide other nutrients found in seafood, meats, and poultry, such as iron and zinc [21]. Indeed, legumes are a critical and affordable source of plant-based proteins, vitamins, and essential minerals such as calcium, magnesium, and zinc, contributing to the food security and nutrition of people around the world, especially subsistence smallholder farmers in developing countries [6]. Human and soil zinc deficiency can be correlated in Africa, Asia, Andean South America, and Mexico [22–24]. In these same regions, where beans are common sources of protein, beans can also be an excellent source of zinc for human diets [5,21]. Bean consumption in sufficient quantities is considered a strategic remedy for hidden hunger and healthy diets [25].

The domestication of common beans was unusual, occurring in two distinct geographic regions simultaneously, creating two gene pools, namely, the Middle American and the Andean [26]. The first introduction of common bean from Central/South America into Western Europe most likely took place in the sixteenth century [27,28]. In Italy, beans were introduced in 1515 [29] and created a long tradition that allowed the evolution of many local varieties adapted to microclimates in restricted areas [30]. In Veneto, the diffusion of common bean occurred quickly, and today, the cultivation of this pulse is of great economic relevance in the Belluno region [31]. However, although those local varieties have original morpho-agronomic and nutritional characteristics, with high organoleptic qualities, they have been gradually substituted by genetically uniform commercial varieties [30,32]. In order to conserve, preserve, and valorize this genetic material, the Department of Agronomy, Food, Natural Resources, Animals, and Environment (DAFNAE) from the University of Padova is establishing a bean germplasm that currently contains 48 accessions, of which 27 are Venetian local varieties, with the preservation of functional and serependic opportunities afforded by plant species diversity for a more sustainable agriculture future.

Most of the world bean production areas are under risk of facing drought conditions during the cultivation cycles [33–36]. These areas are also under high and/or very high risk of human-induced desertification in upcoming years [37], indicating that the effects of drought conditions on bean production may be exacerbated. Drought can reduce bean yields by up to 75% [33,34] or, in more drastic conditions, can lead to total loss of production. At the same time, intensive irrigation systems can lead to an overuse of water and wasting the excess, which may lead to negative impacts on the environment. The long-term objective of the DAFNAE germplasm is to improve Italian and Venetian beans to be used in sustainable organic farming systems. Thus, it is important to better understand the plant development and agronomic performances of some of these local varieties under different environments and irrigation regimes to develop management strategies that can stimulate and support farmers to adopt these beans in their production systems. In addition, it is also important assess their nutraceutical properties in order to valorize this genetic material.

This study aimed to assess the effect of drought conditions during different growth stages of six local bean varieties traditionally cultivated in Veneto (Italy) on agronomic performance and nutraceutical characteristics. Fasolo del Diávolo, Gialet, Posenati, Secle, D'oro, and Maron were selected from the DAFNAE/UNIPD bean germplasm.

2. Materials and Methods

To ensure that the plant responses were primarily related to the drought stress, all the treatments applied were not restricted in terms of environmental factors (i.e., temperature, relative humidity, fertilization, operational practices), which were exactly the same for all cultivated plants, except the irrigation regime. Four different irrigation regimes were imposed on these local varieties: never stressed, always stressed (from sowing to harvesting), stressed before flowering, and stressed after flowering. The irrigation regimes were controlled by using a precipitation exclusion rain-out shelter and the irrigation was applied based on sensors in the pots measuring the volumetric water content every hour. The effect of these different irrigation regimes was assessed on the yield and seeds' antioxidant capacity, total phenols, phosphorus, zinc, calcium, magnesium, and protein content.

2.1. Plant Materials

The Department of Agronomy, Food, Natural Resources, Animals, and Environment (DAFNAE) from the University of Padova (UNIPD), located in Veneto (Italy), is establishing a bean germplasm aiming to conserve beans traditionally cultivated in Italy and, more specifically, in Veneto. This germplasm is composed of 48 accessions and, for this study, we selected six accessions that were tradionally cultivated for centuries by small farmers in the mountainous region of Veneto and selected by them based on their agronomic performance, visual traits, and nutraceutical porperties. Fasolo del Diavolo (Devil's Bean) is a *Phaseolus coccineus*, also known as a runner bean. Gialet (yellowish) and Posenati (from Posina, Veneto) are *P. vulgaris* from the Middle American domestication center and Secle, D'oro (made of gold), and Maron (brown) are *P. vulgaris* from the Andean domestication center (unpublished data) (Figure 1).

Figure 1. Seeds of landraces used in this study, showing the high diversity of visual traits and potential for commercial interests.

2.2. Experimental Design

A pot study was conducted in a rain-out shelter located in the state of Georgia, USA (33°46′53.8248″ N, 83°19′49.7028″ W) from March 2020 until August 2020. The rain-out shelter was completely covered on rainy days. On sunny and cloudy days, the sides of the rain-out shelter were removed, allowing air to flow through it. Each landrace was subjected to four treatments: never stressed (NES), always stressed (ALS), stressed before flowering (SBF), and stressed after flowering (SAF). From sowing until flowering, NES and SAF received a normal irrigation regime and ALS and SBF received a reduced irrigation regime. After flowering, NES and SBF received a normal irrigation regime, whereas ALS and SAF received a reduced irrigation regime. Pots were arranged in a randomized complete block design with four replications of each treatment. A Decagon Em50 data logger was placed in the middle of the shelter and connected to a VP-4 Humidity/Temperature/Barometer sensor (Meter Group Inc., Pullman, WA, USA)) to monitor and record air temperature and relative humidity every hour. The temperature inside the rain-out shelter ranged from 1.2 (2 April at 8 a.m.) to 38.5 °C (4 July at 3 p.m.), and the average temperature was 22.6 °C ± 8.3 (Figure 2a). Four 5TM Moisture/Temperature soil probes were placed at a 5 cm depth into different treatment pots to monitor hourly soil water content. Four reference pots without plants were placed randomly and weighed before and after each irrigation to monitor water evaporation. After the study was concluded, sensors were calibrated in the lab using the same soil with different water contents. Gravimetric water content was determined and converted to volumetric water content using the soil density. Unprocessed data were converted into volumetric water content (VWC, m^3 of water per m^3 of soil) using a calibration curve (R^2 = 0.9342), and the hourly average volume of water per pot of each treatment before and after flowering was calculated. The field capacity (FC) and the permanent wilting point (PWP) of this soil were 46 and 16% (VWC), respectively, or 1.62 and 0.56 L of water per pot. Plant available water was estimated as the difference between water content and permanent wilting point (Figure 2b).

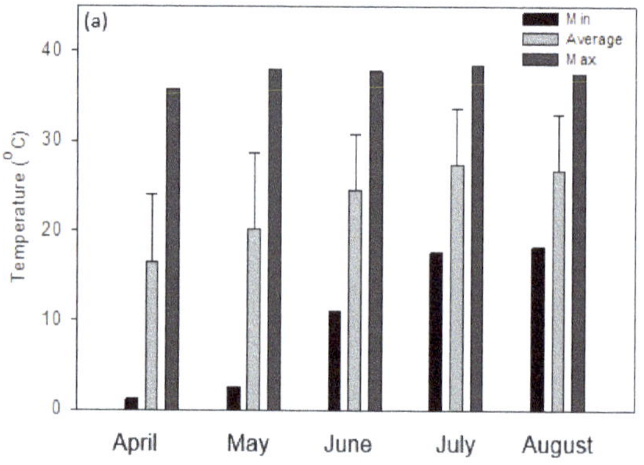

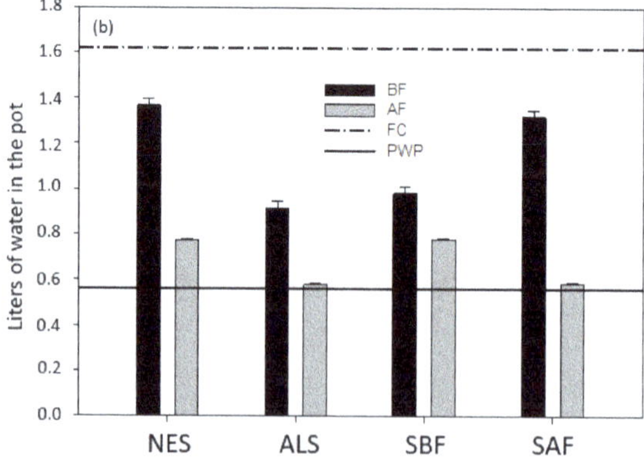

Figure 2. (**a**) Minimum, maximum, and average temperature inside the rain-out shelter from April 2020 until August 2020. (**b**) Hourly average of liters of water per pot in each treatment before flowering (black, BF) and after flowering (gray, AF). Field capacity (FC) is represented by the dash-dot line and permanent wilting point is represented by the solid line. Plant available water is the difference between the volume of water in the pot and the permanent wilting point. Treatments were (1) never stressed (NES), (2) always stressed (ALS), (3) stressed before flowering (SBF), and (4) stressed after flowering (SAF). The standard deviation is represented by error bars.

Each pot with a 20-cm diameter and 3.8 L volume was filled with 3 kg of certified organic soil compost (Soil Cube, Greenville, SC, USA) and two seeds were sowed on 30 March 2020, followed by 500 mL of water to saturate the soil. The soil used had a high organic matter content, 32% loss-on-ignition (LOI) [38], pH 6.8 in 0.01 M $CaCl_2$ (1:2) and 7.3 in distilled H_2O (1:2), bulk density of 0.85 g cm^{-3}, 22.8 mg N kg^{-1} of potentially mineralizable nitrogen by hot KCl methods, 57.5 mg kg^{-1} of plant available N, and CEC of 22.8 meq 100 g^{-1}. The C and N contents of the soil were 13.68% and 1.12%, respectively, which correspond to a C/N ratio of 12.2. Other nutrient contents were 4,845 mg Ca kg^{-1}, 1808 mg K kg^{-1}, 1270 mg Mg kg^{-1}, 632 P mg kg^{-1}, and 58 mg Zn kg^{-1}. Five days after planting (DAP), all plants emerged, and at three days after emergence, one plant was thinned from each pot, ensuring that all plants within each landrace were as homogenous as possible. All plants bloomed between 35 and 52 days after planting (DAP).

2.3. Plant Development and Yield

All five *P. vulgaris* have an indeterminate IIIb growth habit and *P. coccineus* IVa, according to the CIAT scale [39]. We used one cotton twine for each plant to support growth. One end was tied to the top of the rain-out shelter and the other end to the bottom of the plant. In the first weeks, we wrapped the stem around the twine, and the plants utilized the twine.

Every ten days, each plant development stage was monitored following the bean BBCH scale proposed by Feller et al. (1995) [40]. This scale has nine different growth stages, but for this study, the scale was adapted and summarized into four main growth stages:

i.   Vegetative phase: from germination until first flower opens;
ii.   Flowering: from the first flowers opening until the first pods are visible;
iii. Fresh pods: from the first green pod being visible until the first pod starts to dry and ripen;
iv.   Dry pods: from when the pods and seeds ripen until the plant's senescence.

Pods were harvested when they appeared dry, and then seeds were left in a room with low relative humidity and temperature of about 20 °C, weighed, and the dry seed production per plant was determined.

2.4. Seeds' Nutraceutical Characterization

Air-dried seeds were ground to fine powder at room temperature, and approximately 0.2 g of each sample was added to a 50-mL Falcon tube and homogenized in 20 mL of methanol. Samples were filtered in Whatman 4 grade filter paper, and appropriate aliquots of extracts were assayed by Folin–Ciocalteau assay for total phenol content and by Ferric Reducing Antioxidant Power assay for total antioxidant activity. For all the nutraceutical properties analyzed, we used three technical replications for each biological replication.

The content of total phenols was measured on the basis of mg of gallic acid equivalent per kg (mg GAE kg^{-1} dw) of dry bean powder [41]. Bean extract was mixed with 20% sodium carbonate solution and Folin–Ciocalteau reagent [42]. After 2 h at room temperature, the absorbance of the colored reaction product was measured at 765 nm. The evaluation of total phenols was conducted utilizing a standard calibration curve (R^2 = 0.9995) of gallic acid solution with concentrations from 0 to 1600 μg mL^{-1} of gallic acid equivalent.

FRAP reagent was prepared fresh so that it contained 1 mM 2,4,6 tripyridyl-2-triazine and 2 mM ferric chloride in 0.25 M sodium acetate at pH 3.6 [43]. A 100-μL aliquot of the methanol extract prepared as above was added to 1900 μL of FRAP reagent and accurately mixed. After leaving the mixture at 20 °C for 4 min, the absorbance at 593 nm was determined. The evaluation of antioxidant capacity was conducted utilizing a standard calibration curve (R^2 = 0.9989) of ferrous ammonium with concentrations from 0 to 1200 μg mL^{-1} of Fe2+E (ferrous ion equivalent).

Approximately 0.2 g of dried ground bean powder was digested on two cycles of 200 °C for one hour and at 375 °C for two and a half hours, with 3 mL sulfuric acid being added at each cycle. Catalyst was added in the first cycle only. Total protein was calculated from total nitrogen measured from the Kjeldhal digestion [44], and total phosphorus was determined using the method described by Jirka et al. (1976) [45]. Calcium, zinc, and magnesium were measured in a Perkin Elmer AAnalyst 200 atomic absorption unit using standard curves obtained by diluting 1000 ppm stock solutions (Fischer Chemical) in 2% nitric acid.

2.5. Statistical Analysis

All data collected were digitized and processed on Microsoft Excel. A summary spreadsheet with the data of each variable of each experimental unit was made and input into JMP Pro 15 (JMP, version Pro 15, SAS Institute Inc., Cary, NC, USA, 1989–2019) that was used to perform the statistical analysis. The data were assessed for normality and tested for homogeneity in JMP Pro 15 and found to be normally distributed and homogenous. Analysis of variance (ANOVA) considering the main effects of treatment, variety, and their interaction was performed, and the Tukey test was used to separate means at $p < 0.05$. Principal component analysis (PCA) and the correlation between the variables were also performed. SigmaPlot 11.0 (Systat Software, San Jose, CA, USA) was used to prepare graphs.

3. Results

Analysis of variance and interactions in Table 1 show the incidence of interactions in some measured parameters. The structure of the presented Results and Discussion follows the order in which the parameters are presented in Table 1.

Table 1. Analysis of variance (ANOVA) table with the effects on different local varieties, of treatments, and of the interaction of variety and treatments on yield and nutraceutical properties of Venetian bean (*P. vulgaris*) local varieties cultivated in a rain-out shelter between March and August 2020, except Fasolo del Diavolo that did not produce seeds. The *p*-values are followed by the F-values in parenthesis.

Parameters	Varieties	Treatments	Varieties × Treatments
Yield (g per plant)	<0.001 (11.71)	<0.001 (14.57)	0.021 (2.56)
Total phenols (mg GAE kg^{-1} dw)	<0.001 (17.13)	0.065 (2.87)	0.798 (0.63)
Antioxidant capacity (mg Fe^{+2} kg^{-1} dw)	0.006 (4.57)	0.301 (1.22)	0.296 (1.21)
Protein content (%)	0.198 (1.64)	0.057 (3.10)	0.813 (0.60)
Total phosphorus content (mg kg^{-1})	0.196 (1.69)	0.698 (0.36)	0.053 (2.21)
Mg content (mg kg^{-1})	0.231 (1.49)	0.489 (0.73)	0.597 (0.85)
Ca content (mg kg^{-1})	<0.001 (8.72)	0.111 (2.32)	0.149 (1.43)
Zn content (mg kg^{-1})	<0.001 (30.43)	0.070 (2.83)	0.002 (4.33)
Degrees of freedom	4	3	12

Total degrees of freedom = 79.

3.1. Yield

The interaction between local varieties and treatment had a significant effect on yield (Table 1). Posenati NES and SBF and Secle SBF produced more than 10 g of seeds per plant. Posenati NES and SBF produced significantly more than Posenati SAF and ALS. Secle SBF yield was significantly greater than either Secle SAF or SBF. Gialet was the only local variety that did not have its yield significantly affected when comparing the treatments within it. D'oro SAF yield was significantly the lowest and Maron ALS did not produce seeds (Figure 3). Fasolo del Diavolo did not develop pods and therefore it did not produce seeds.

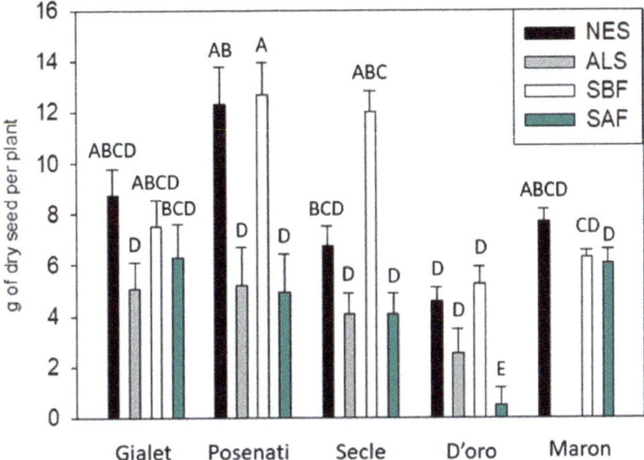

Figure 3. Effects of varieties and treatments interaction on yield (g of dry seeds per plant) of Venetian bean (*P. vulgaris*) local varieties cultivated in a rain-out shelter between March and August 2020. Different letters indicate a significant difference between compared groups. α = 0.05. Treatments were (1) never stressed (NES), (2) always stressed (ALS), (3) stressed before flowering (SBF), and (4) stressed after flowering (SAF). Maron ALS did not produce seeds; thus, it is not represented in the graph. The standard deviation is represented by error bars (n = 4).

3.2. Nutraceutical Properties

Total phenols content was significantly affected by variety (Table 1). Maron and D'oro significantly had the greatest total phenols content, whereas Gialet significantly had the lowest (Table 2). The antioxidant capacity was also significantly affected by variety (Table 1). Maron had the greatest antioxidant capacity and Gialet the lowest, with a non-significant difference when compared to Secle (Table 2). Local varieties significantly affected the calcium content on the seeds (Table 1), with Gialet and Posenati having a significantly greater Ca content than Secle and D'oro (Table 2).

Table 2. Seeds' nutraceutical characteristics of five Venetian bean (*P. vulgaris*) local varieties cultivated in a rain-out shelter between March and August 2020.

Parameters	Gialet	Posenati	Secle	D'oro	Maron
Total phenols (mg GAE Kg^{-1} dw)	2455 (76) c	3368 (76) b	3326 (52) b	4288 (77) a	4128 (67) a
Antioxidant capacity (mg Fe^{+2} kg^{-1} dw)	2505 (36) c	3718 (85) b	3256 (88) bc	3736 (154) b	4545 (135) a
Protein content (% dw) [n.s.]	30.9 (4.4)	30.3 (2.4)	28.2 (3.4)	32.3 (2.8)	27.1 (3.5)
Total phosphorus (mg kg^{-1} dw) [n.s.]	1395 (36)	1386 (54)	1439 (81)	1422 (28)	1379 (55)
Mg content (mg kg^{-1} dw) [n.s.]	297 (13.16)	287.8 (14.7)	292.7 (12.4)	296.3 (6.5)	281.3 (11.1)
Ca content (mg kg^{-1} dw)	1280 (163) a	1118 (193) a	996 (194) b	945 (133) b	1081 (132) ab

Means followed by different letters in the same row indicate a significant difference between compared local varieties. n.s. = non-significant difference. α = 0.05. The standard deviation is in parenthesis after the values (n = 4). Maron ALS did not produce seeds; thus, it is not represented in the table.

Zinc content was significantly affected by the interaction between variety and treatment (Table 1). Gialet SBF and SAF and Maron NES had Zn content greater than 60 mg kg^{-1}, whereas Posenati SBF and SAF and D'oro SAF had Zn content lower than 40 mg kg^{-1}. On average, Gialet and Maron zinc content was about 60 mg kg^{-1} (Table 3).

Table 3. Effects of varieties and treatment interaction on zinc content of Venetian bean (*P. vulgaris*) local varieties cultivated in a rain-out shelter between March and August 2020. Treatments were (1) never stressed (NES), (2) always stressed (ALS), (3) stressed before flowering (SBF), and (4) stressed after flowering (SAF).

Varieties	Treatment	Zn Content (mg kg^{-1})
Gialet	NES	58.4 (1.9) abcd
	ALS	56.8 (2.3) abcde
	SBF	61.8 (6.8) abc
	SAF	65.6 (6.7) ab
Posenati	NES	43.9 (2.7) efg
	ALS	47.4 (3.6) defg
	SBF	37.1 (5.2) g
	SAF	38.7 (2.4) fg
Secle	NES	47.9 (3.0) cdefg
	ALS	49.2 (1.8) bcdefg
	SBF	58.3 (8.6) abcd
	SAF	50.9 (4.8) bcdefg
D'oro	NES	49.4 (0.3) bcdefg
	ALS	56.2 (1.1) abcdef
	SBF	41.2 (1.6) fg
	SAF	37.5 (2.7) fg
Maron	NES	69.2 (0.3) a
	ALS	—
	SBF	58.3 (7.1) abcd
	SAF	51.4 (1.8) abcdefg

Means followed by different letters indicate a significant difference between compared groups. $\alpha = 0.05$. The standard deviation is in parenthesis after the values. Maron ALS did not produce seeds; thus, it is not represented in the table.

3.3. Principal Component Analysis and Correlations

Considering the relationship of yield with nutritional and nutraceutical parameters, a principal component analysis on covariance was performed. The total variation of C1 + C2 represented 98.4% of the total variability. Ca, Zn, and yield are on the C1 negative and C2 negative quarters, whereas total phenols and protein are on the C1 positive and C2 positive quarters. Antioxidant capacity is on the C1 positive and C2 negative quarters, and Mg and P are on the C1 negative and C2 positive quarters (Figure 4).

Yield correlation was significant ($p < 0.05$) and positive with calcium and siginificant and negative with total phenols, antioxidant capacity, and protein content. Total phenols and protein content were significant and positively correlated. Total phenols was significant and negatively correlated with zinc and calcium. Antioxidant capacity also had a significant and negative correlation with calcium and zinc (Table 4).

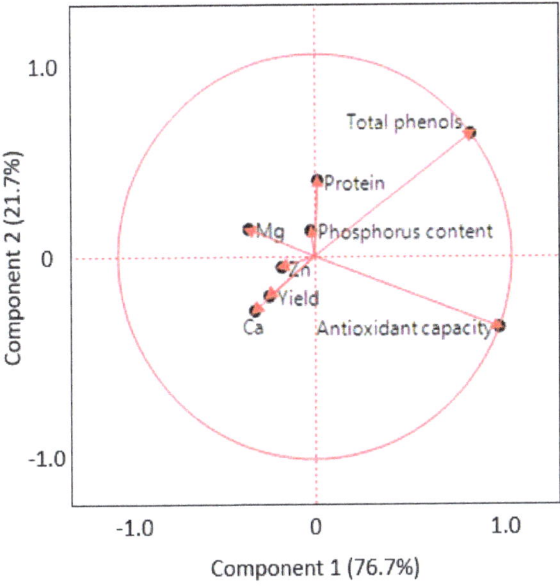

Figure 4. Principal component analysis (PCA) on covariance of nutraceutical properties of Venetian bean (*P. vulgaris*) local varieties cultivated in a rain-out shelter between March and August 2020.

Table 4. Correlations coefficients of the variables of Venetian bean (*P. vulgaris*) local varieties cultivated in a rain-out shelter between March and August 2020, estimated by the restricted maximum likelihood (REML) method.

	Yield	TP	AC	Protein	P	Zn	Mg	Ca
Yield	1							
TP	−0.5354 *	1						
AC	−0.3431 *	0.6675	1					
Protein	−0.2955 *	0.3749 *	−0.0078	1				
P	−0.1147	0.1566	0.0458	−0.1097	1			
Zn	0.0198	−0.2747 *	−0.2994 *	−0.1471	−0.0765	1		
Mg	0.0019	−0.1941	−0.2381	−0.0455	0.0997	−0.0504	1	
Ca	0.5062 *	−0.6477 *	−0.4166 *	−0.1714	−0.1023	0.2366	0.2335	1

TP = total phenols; AC = antioxidant capacity. * significant correlation ($p < 0.05$).

4. Discussion

This study assessed the effect of drought conditions at different growing stages on the yield and nutraceutical properties of six beans traditionally cultivated in Veneto, Italy. The local varieties had different responses to different irrigation regimes. Gialet was the only one that was not significantly affected by drought conditions and had a high zinc content. We also found that yield was significant and negatively correlated with protein content, total phenols, and antioxidant capacity. In this study, calcium was significant and positively correlated with yield.

Fasolo del Diavolo did not develop pods in this experiment, possibly due to two reasons: (i) *P. coccineus* has a high degree of allogamy, needing the presence of specific pollinators. We speculate that the rain-out shelter was not conducive to the presence of these pollinators or conditions within the rain-out shelter resulted in the reduced incidence of proper plant–pollinator contact [46]. (ii) *P. coccineus* tolerates temperatures up to 30 °C, and in this experiment, the temperature inside the rain-out shelter was commonly greater than 35 °C during the daytime. Thus, to assess the effects of drought conditions on this local variety, further research in an environment with a cooler temperature would be recommended. This also indicates that the cultivation of this local variety should be limited to the mountainous environment, where it originates, with milder temperatures throughout the growing period.

Polania et al.'s (2016) results showed that Middle American accessions tend to be more tolerant to drought conditions. The superior performance under drought stress of these accessions was associated with a better canopy biomass at mid-pod filling, which can be related to a deeper root system and effective use of water, an efficient remobilization of photosynthates, and grain filling [47]. Dipp et al.'s (2017) results confirmed that Andean accessions have a higher susceptibility to drought conditions [48]. In this study, Gialet and Posenati represented the Middle American gene pool. Drought conditions did not have significant effects on Gialet's yield, and Posenati's yield was not affected when it was under the SBF treatment. Maron and D'oro, two of the Andean local varieties, had higher susceptibility to drought stress; however, they had different responses. Maron ALS plants were more affected, whereas D'oro's SAF plants were more susceptible. Dipp et al. (2017) found that highly susceptible accessions can have yields reduced by about 70% and they considered that accessions with yield reduction lower than 40% are tolerant to drought conditions [48]. Smith et al. (2019) also found that drought conditions can significantly reduce yield by 70%. The yields of Gialet and Secle, for example, were reduced by 42 and 40%, respectively, under drought conditions during the whole cycle, indicating that these accessions can be tolerant to drought conditions [49].

Stressing the plant before flowering and applying a normal irrigation regime during the reproductive stage increased the yield of Posenati and D'oro and significantly increased Secle's yields by 80%. Imposing stress before flowering on beans' cultivation is a strategy recommended for Italian bean local varieties and commercial varieties [50–52]. Thus, irrigation management is a tool that can be used by farmers cultivating these local varieties, reducing the amount of water applied and increasing the seed production and the water use efficiency, without affecting the nutraceutical properties of the seeds.

Smith et al. (2019) also evaluated the effect of drought conditions on the macronutrient and micronutrient contents of bean seeds. Drought conditions did not affect the content of phosphorus, calcium, zinc, and magnesium [49], as was the case in our study. In this study, however, variety and treatments interaction had a significant effect on seed zinc content. Posenati and D'oro ALS seeds had significantly higher content than SBF. Nicoletto et al. (2019) assessed the effect of two different environments (30 and 351 m altitudes) on seed qualities of two Venetian local varieties, Gialet and Lingua di Fuoco, and their results showed that the environment significantly affected antioxidant capacity, total phenols, and calcium, magnesium, and phosphate content in seeds [53]. Although the growth environment affected these parameters, drought conditions did not have significant effects on seeds' nutraceutical characteristics. However, there was a significant difference in nutraceutical properties among the local varieties. D'oro and Maron had significantly greater total phenols content and Maron also had the greatest antioxidant capacity, whereas Gialet had significantly the lowest content of total phenols and antioxidant capacity among the local varieties. Most of the antioxidants and total phenols of beans are present on the seed coat and colored seeds are related to high contents of these compounds [14–16], confirming the results of this study, as Gialet seeds are yellowish and pale, whereas D'oro and Maron have strong golden and brownish colors, respectively. In this study, local varieties from the Andean domestication center tended to have, on average, greater antioxidant capacity and total phenols content and lower calcium content, when compared to Middle American local varieties. Nicoletto et al.'s (2019) results for antioxidant capacity and total phenols of Gialet seeds were lower than the ones obtained in this study [53]. However, these results are based on the weight of dry seeds and previous studies have shown that soaking and cooking beans can reduce by 90% the total phenol content and antioxidant capacity [54–56].

Blair et al. (2009) considered a Middle American accession with 42.8 mg kg^{-1} Zn content on the seeds as a low content and an Andean accession with 66.7 mg kg^{-1} as high

Zn content [57]. Amongi et al. (2018) assessed the Zn content of 304 accessions of a common bean germplasm from East Africa. The Zn content on African accessions ranged from 23.9 to 47 mg kg^{-1}, with an average of 32 mg kg^{-1} [58]. This study obtained a similar range to Blair's, from 42.2 (Posenati) to 59.3 and 59.9 mg Zn kg^{-1} (Maron and Gialet, respectively). These local varieties had greater Zn concentrations' than modern varieties assessed by Celmeli et al. (2018) that ranged from 25 to 35 mg Zn kg^{-1} [59]. The average daily recommended amount of Zn consumption for an adult male is 11 mg [60], which means that if one adult male consumes the equivalent of 100 g of dry beans from a commercial variety containing 32 mg Zn kg^{-1}, it will provide about 29% of the recommended daily consumption of Zn. If the same man eats the same amount of dry beans from a Zn-rich local variety—such as Gialet and Maron, for example—it will provide about 55% of the recommended daily consumption of Zn in one meal. Paredes et al. (2010) assessed the macronutrient and micronutrient content of 52 Chilean bean accessions that had Mg content ranging from 1300 to 1800 mg kg^{-1}, about five times greater than the ones obtained in this study. In contrast, Smith et al. (2019) measured Mg magnesium contents closer to the ones from this study, about 500 mg kg^{-1}. In terms of Ca content, this study's results are closer to those of Paredes et al. (2010), which ranged from 1000 to 2100 mg Ca kg^{-1}, whereas Smith et al. (2019) found Ca contents around 500 mg kg^{-1}. Paredes et al. (2010) used an atomic absorption unit to determine Ca and Mg, as was used in this study, and Smith et al. (2019) used inductively coupled plasma (ICP) [49,61].

Table 4 indicates that the yield correlation with protein was negative (−0.2955) and significant ($p < 0.005$), in agreement with the results of Bulyaba et al. (2020) and Kazai et al. (2019) [62,63]. Davis et al. (2004) proposed a trade-off theory for the interaction between yield and nutrient content [64]. Thus, we hypothesize that, in this study, the trade-off of reduced protein content may be explained by the dilution effect since the increase in yield may result in a greater carbohydrates content and reduced nitrogen content within the endosperm. Total phenols and antioxidant capacity were also negatively correlated with yield for beans. Kruger et al. (2013) found that both were also negatively related to the yield of 23 accessions of black currants [65]. It is likely that greater yields increased the endosperm vs. coat ratio, and since most of the antioxidants and phenols are in the seed coat [14–16], their contents were decreased as yield increased, leading to the negative correlation found in this study. On the contrary, yield had a positive and significant correlation with calcium content (0.5062, $p < 0.05$). Ribeiro et al. (2013) found out that yield was positively correlated with Ca content (0.1566) [66]. However, in another study, Ribeiro et al. found a non-significant negative correlation between yield and calcium content [67]. Thus, we hypothesize that the correlation between yield and calcium content in the seeds may be explained by another factor besides the dilution effect, for example, a high Ca content in the soil, soil pH, different varietal requirements and Ca uptake, seed nutritional characteristics, and response to drought conditions. Domingues et al. (2016) found that a high bioavailability of Ca is correlated with higher yields and higher Ca content in the seeds. Since the soil used in this study had a high Ca content, we hypothesized that these local varieties were able to take advantage of the higher calcium content in the soil [68].

5. Conclusions

Exposure to drought conditions until flowering did not significantly reduce yields for any of the local varieties and instead increased Secle's yield by 80% when compared to the never stressed treatment. Thus, reducing irrigation before flowering is a tool that can be used by farmers cultivating these varieties to increase yield and their water use efficiency. Gialet was the only variety that did not have yields significantly affected by drought conditions for any of the treatments: before flowering, after flowering, and throughout the cultivation cycle. Maron did not produce seeds when exposed to drought conditions throughout the cultivation cycle. Drought conditions during different growing stages did not affect seeds' nutraceutical properties, except for zinc which was significantly affected by the interaction between variety and treatment. Gialet and Maron had a high zinc content when compared to commercial varieties (60 mg kg^{-1}). Gialet, a traditional variety from Veneto, has good potential of success when cultivated in areas under risk of facing drought conditions and is a good candidate for addressing the zinc deficiency faced by a significant proportion of the world's population. While these local varieties have some cultivation limitations for large commercial growers (i.e., indeterminate growth, manual harvesting, and practices to increase their efficiency and productivity are still little known), this study provides useful information to small and medium farmers that will help them to increase the water use efficiency and the nutritional values of these beans, which can be grown in other drought regions of the world. It is also possible that these results will help and promote the commercial value of these varieties and raise the interest of farmers to start producing them on larger scales. However, more studies in different environments under drought conditions are still needed in order to provide more information to bean growers.

Author Contributions: Conceptualization, P.S. (Pietro Sica), C.P., M.C., D.F., A.G., F.S., C.M., C.N., P.S. (Paolo Sambo), G.B. and M.B.; formal analysis, P.S. (Pietro Sica), M.C. and D.F.; methodology, P.S. (Pietro Sica), C.P., M.C. and D.F; investigation, P.S. (Pietro Sica), M.C. and D.F.; resources, M.C. and D.F.; writing—original draft preparation, P.S. (Pietro Sica); writing—review and editing, P.S. (Pietro Sica), M.C., D.F., M.B. and G.B.; visualization, P.S. (Pietro Sica), M.C. and D.F.; supervision, M.C., D.F., G.B. and M.B.; project administration, M.C., D.F., P.S. (Paolo Sambo), C.N., G.B. and M.B. All authors have read and agreed to the published version of the manuscript.

Funding: This research received no external funding. "The APC was funded by the University of Georgia, Department of Crop & Soil Sciences".

Institutional Review Board Statement: Not applicable.

Informed Consent Statement: Not applicable.

Data Availability Statement: The data presented in this study are available on request from the corresponding author. The data are not publicly available due to privacy.

Acknowledgments: The Venetian landraces of common bean used in this study are conserved at the Germplasm Bank of DAFNAE: seed and plant materials of this crop have been collected, characterized and pre-selected within the activities run for the VALEBIO project funded by the University of Padova (UNI-IMPRESA call 2018).

Conflicts of Interest: The authors declare no conflict of interest.

References

1. FAO. Pulses Contribute to Food Security. Available online: http://www.fao.org/fileadmin/user_upload/pulses-2016/docs/factsheets/FoodSecurity_EN_PRINT.pdf (accessed on 30 August 2020).
2. Jimenez-Lopez, J.C.; Singh, K.B.; Clemente, A.; Nelson, M.N.; Ochatt, S.; Smith, P.M.C. Editorial: Legumes for Global Food Security. *Front. Plant. Sci.* **2020**, *11*, 926. [CrossRef] [PubMed]
3. Blair, M.W. Mineral biofortification strategies for food staples: The example of common bean. *J. Agric. Food Chem.* **2013**, *61*, 8287–8294. [CrossRef] [PubMed]
4. McClean, P.E.; Raatz, B. Common Bean Genomes: Mining New Knowledge of a Major Societal Crop. In *The Common Bean Genome*; de la Vega, M.P., Marsolais, F., Santalla, M., Eds.; Springer International Printer: Gewebestrasse, Switzerland, 2017; p. 295. ISBN 978-3-319-63526-2.
5. Jones, A. Phaseoulus Beans: Post-Harvest Operations. Available online: http://www.fao.org/3/a-av015e.pdf (accessed on 13 January 2020).
6. Calles, T. Preface to special issue on leguminous pulses. *Plant. Cell Tiss Organ. Cult.* **2016**, *127*, 541–542. [CrossRef]
7. Peix, A.; Ramírez-Bahena, M.H.; Velázquez, E.; Bedmar, E.J. Bacterial Associations with Legumes. *CRC Crit. Rev. Plant. Sci.* **2015**, *34*, 17–42. [CrossRef]
8. Courty, P.E.; Smith, P.; Koegel, S.; Redecker, D.; Wipf, D. Inorganic Nitrogen Uptake and Transport in Beneficial Plant Root-Microbe Interactions. *CRC Crit. Rev. Plant. Sci.* **2015**, *34*, 4–16. [CrossRef]
9. FAO. *How to Feed the World in 2050*; FAO: Rome, Italy, 2009.
10. Arnoldi, A.; Zanoni, C.; Lammi, C.; Boschin, G. The Role of Grain Legumes in the Prevention of Hypercholesterolemia and Hypertension. *CRC Crit. Rev. Plant. Sci.* **2015**, *34*, 144–168. [CrossRef]
11. Guzmán-Maldonado, S.H.; Acosta-Gallegos, J.; Paredes-López, O. Protein and mineral content of a novel collection of wild and weedy common bean (*Phaseolus vulgaris* L). *J. Sci. Food Agric.* **2000**, *80*, 1874–1881. [CrossRef]

12. Mitchell, C.J.; Milan, A.M.; Mitchell, S.M.; Zeng, N.; Ramzan, F.; Sharma, P.; Knowles, S.O.; Roy, N.C.; Sjödin, A.; Wagner, K.-H.; et al. The effects of dietary protein intake on appendicular lean mass and muscle function in elderly men: A 10-wk randomized controlled trial. *Am. J. Clin. Nutr.* **2017**, *106*, 1375–1383. [CrossRef]

13. Rand, W.M.; Pellett, P.L.; Young, V.R. Meta-analysis of nitrogen balance studies for estimating protein requirements in healthy adults. *Am. J. Clin. Nutr.* **2003**, *77*, 109–127. [CrossRef]

14. Díaz-Batalla, L.; Widholm, J.M.; Fahey, G.C.; Castaño-Tostado, E.; Paredes-López, O. Chemical Components with Health Implications in Wild and Cultivated Mexican Common Bean Seeds (*Phaseolus vulgaris* L.). *J. Agric. Food Chem.* **2006**, *54*, 2045–2052. [CrossRef]

15. Ombra, M.N.; D'acierno, A.; Nazzaro, F.; Riccardi, R.; Spigno, P.; Zaccardelli, M.; Pane, C.; Maione, M.; Fratianni, F. Phenolic Composition and Antioxidant and Antiproliferative Activities of the Extracts of Twelve Common Bean (*Phaseolus vulgaris* L.) Endemic Ecotypes of Southern Italy before and after Cooking. *Oxid. Med. Cell. Longev.* **2016**, *2016*, 1398298. [CrossRef] [PubMed]

16. Oroian, M.; Escriche, I. Antioxidants: Characterization, natural sources, extraction and analysis. *Food Res. Int.* **2015**, *74*, 10–36. [CrossRef] [PubMed]

17. Cardador-Martínez, A.; Albores, A.; Bah, M.; Calderón-Salinas, V.; Castaño-Tostado, E.; Guevara-González, R.; Shimada-Miyasaka, A.; Loarca-Piña, G. Relationship among antimutagenic, antioxidant and enzymatic activities of methanolic extract from common beans (*Phaseolus vulgaris* L). *Plant. Foods Hum. Nutr.* **2006**, *61*, 161–168. [CrossRef] [PubMed]

18. Kalogeropoulos, N.; Chiou, A.; Ioannou, M.; Karathanos, V.T.; Hassapidou, M.; Andrikopoulos, N.K. Nutritional evaluation and bioactive microconstituents (phytosterols, tocopherols, polyphenols, triterpenic acids) in cooked dry legumes usually consumed in the Mediterranean countries. *Food Chem.* **2010**, *121*, 682–690. [CrossRef]

19. Ranilla, L.G.; Genovese, M.I.; Lajolo, F.M. Effect of different cooking conditions on phenolic compounds and antioxidant capacity of some selected Brazilian bean (*Phaseolus vulgaris* L.) cultivars. *J. Agric. Food Chem.* **2009**, *57*, 5734–5742. [CrossRef]

20. NIH–U.S. (National Center for Complementary and Integrative Health) Antioxidants: In Depth | NCCIH. Available online: https://www.nccih.nih.gov/health/antioxidants-in-depth (accessed on 7 July 2020).

21. USDA (U.S. Department of Agriculture) and HHS (U.S. Department of Health and Human Services). 2015–2020 Dietary Guidelines for Americans, 8th Edition. Available online: https://health.gov/our-work/food-nutrition/previous-dietary-guidelines/2015 (accessed on 30 August 2020).

22. Alloway, B.J. *Zinc In Soils and Crop Nutrition*, 2nd ed.; IZA and IFA: Paris, France, 2008; ISBN 978-90-8133-310-8.

23. Wessells, K.R.; Brown, K.H. Estimating the Global Prevalence of Zinc Deficiency: Results Based on Zinc Availability in National Food Supplies and the Prevalence of Stunting. *PLoS ONE* **2012**, *7*, e50568. [CrossRef]

24. Cakmak, I.; McLaughlin, M.J.; White, P. Zinc for better crop production and human health. *Plant. Soil* **2017**, *411*, 1–4. [CrossRef]

25. Larochelle, C.; Katungi, E.; Beebe, S. *Disaggregated Analysis of Bean Consumption Demand and Contribution to Household Food Security in Uganda*; CGIAR: Cali, Colombia, 2015.

26. Bitocchi, E.; Rau, D.; Bellucci, E.; Rodriguez, M.; Murgia, M.L.; Gioia, T.; Santo, D.; Nanni, L.; Attene, G.; Papa, R. Beans (Phaseolus ssp.) as a model for understanding crop evolution. *Front. Plant. Sci.* **2017**, *8*. [CrossRef]

27. Zeven, A.C. The introduction of the common bean (*Phaseolus vulgaris* L.) into Western Europe and the phenotypic variation of dry beans collected in the Netherlands in 1946. *Euphytica* **1997**, *94*, 319–328. [CrossRef]

28. De Ron, A.M.; González, A.M.; Rodiño, A.P.; Santalla, M.; Godoy, L.; Papa, R. History of the common bean crop: Its evolution beyond its areas of origin and domestication. *Arbor* **2016**, *192*, a317. [CrossRef]

29. Albala, K. Phaseolus vulgaris: Mexico and the World. In *Beans: A History*; Bloomsbury Academic: London, UK, 2007; p. 240. ISBN 978-1-8452-0430-3.

30. Venora, G.; Grillo, O.; Ravalli, C.; Cremonini, R. Identification of Italian landraces of bean (*Phaseolus vulgaris* L.) using an image analysis system. *Sci. Hortic.* **2009**, *121*, 410–418. [CrossRef]

31. Piergiovanni, A.R.; Lioi, L. Italian Common Bean Landraces: History, Genetic Diversity and Seed Quality. *Diversity* **2010**, *2*, 837–862. [CrossRef]

32. Spagnoletti Zeuli, P.L.; Baser, N.; Riluca, M.; Laghetti, G.; Logozzo, G.; Masi, P.; Molinari, S.; Negri, V.; Olita, G.; Tiranti, B.; et al. Valorisation and certification of Italian bean agro-ecotypes (*Phaseolus vulgaris*). In Proceedings of the Ecotipi Vegetali Italiani: Una Preziosa Risorsa di Variabilità Genetica, Roma, Italy, 7 October 2004; p. 19. (In Italian).

33. Beebe, S.E.; Rao, I.M.; Blair, M.W.; Acosta-Gallegos, J.A. Phenotyping common beans for adaptation to drought. *Front. Physiol.* **2013**, *4*, 35. [CrossRef] [PubMed]

34. Broughton, W.J.; Hernández, G.; Blair, M.; Beebe, S.; Gepts, P.; Vanderleyden, J. Beans (*Phaseolus* spp.)—Model Food Legumes. *Plant Soil* **2003**, *252*, 55–128. [CrossRef]

35. FAOSTAT Beans, Dry. Available online: http://www.fao.org/faostat/en/#data/QC (accessed on 7 January 2020).

36. NIDIS Current Conditions | Global Drought Information System. Available online: https://www.drought.gov/gdm/current-conditions (accessed on 4 June 2020).

37. USDA Risk of Human-Induced Desertification Map | NRCS Soils. Available online: https://www.nrcs.usda.gov/wps/portal/nrcs/detail/soils/survey/geo/?cid=nrcs142p2_054004 (accessed on 9 October 2020).

38. Hendricks, T.; Franklin, D.; Dahal, S.; Hancock, D.; Stewart, L.; Cabrera, M.; Hawkins, G. Soil carbon and bulk density distribution within 10 Southern Piedmont grazing systems. *J. Soil Water Conserv.* **2019**, *74*, 323–333. [CrossRef]

39. Singh, S.P. *A Key for the Identification of Different Growth Habits of Phaseolus vulgaris*; CIAT: Cali, Colombia, 1981.

40. Feller, C.H.; Bleiholder, L.; Buhr, H.; Hack, M.; Hess, R.; Klose, U.; Meier, R.; Stauss, T.; Van Den Boom, E. Phenological stages of development of vegetables: II. Fruit vegetables and legumes. In *Growth Stages of Mono-and Dicotyledonous Plants BBCH Monograph Federal Biological Research Centre for Agriculture and Forestry (2001)*; Meier, U., Ed.; BBA: Berlin, Germany, 1995; pp. 141–144.

41. Singleton, V.L.; Orthofer, R.; Lamuela-Raventós, R.M. Analysis of total phenols and other oxidation substrates and antioxidants by means of folin-ciocalteu reagent. *Methods Enzymol.* **1999**, *299*, 152–178. [CrossRef]

42. Fischer Scientific Folin and Ciocalteu's Phenol Reagent, 2.0N, MP Biomedicals 500 mL: Chemicals | Fisher Scientific. Available online: https://www.fishersci.com/shop/products/folin-ciocalteu-s-phenol-reagent-2-0n-mp-biomedicals-500mL/icn19518690 (accessed on 5 September 2020).

43. Benzie, I.F.F.; Strain, J.J. The ferric reducing ability of plasma (FRAP) as a measure of "antioxidant power": The FRAP assay. *Anal. Biochem.* **1996**, *239*, 70–76. [CrossRef]

44. Hjalmarsson, S.; Akesson, R. Modern Kjeldahl Procedure. *Int. Lab.* **1983**, *3*, 70–76.

45. Jirka, A.M.; Carter, M.J.; May, D.; Fuller, F.D. Ultramicro Semiautomated Method for Simultaneous Determination of Total Phosphorus and Total Kjeldahl Nitrogen in Wastewaters. *Environ. Sci. Technol.* **1976**, *10*, 1038–1044. [CrossRef]

46. Giurcă, D.M. Morphological and phenological differences between the two species of the Phaselous genus (Phaseolus vulgaris and Phaseolus Coccineus). *Cercet. Agron. în Mold.* **2009**, *XLII*, 39–45.

47. Polania, J.; Rao, I.M.; Cajiao, C.; Rivera, M.; Raatz, B.; Beebe, S. Physiological traits associated with drought resistance in Andean and Mesoamerican genotypes of common bean (*Phaseolus vulgaris* L.). *Euphytica* **2016**, *210*, 17–29. [CrossRef]

48. Dipp, C.C.; Marchese, J.A.; Woyann, L.G.; Bosse, M.A.; Roman, M.H.; Gobatto, D.R.; Paludo, F.; Fedrigo, K.; Kovali, K.K.; Finatto, T. Drought stress tolerance in common bean: What about highly cultivated Brazilian genotypes? *Euphytica* **2017**, *213*, 1–16. [CrossRef]

49. Smith, M.R.; Veneklaas, E.; Polania, J.; Rao, I.M.; Beebe, S.E.; Merchant, A. Field drought conditions impact yield but not nutritional quality of the seed in common bean (*Phaseolus vulgaris* L.). *PLoS ONE* **2019**, *14*, e0217099. [CrossRef] [PubMed]

50. Orto Quando Irrigare le Piante di Fagioli. Available online: https://www.ortodacoltivare.it/domande/irrigare-fagioli.html (accessed on 7 September 2020).

51. Albanesi Coltivazione dei Fagioli—Coltivazione dei Borlotti-Albanesi.it. Available online: https://www.albanesi.it/ambiente/orto/coltivazione-fagioli.htm (accessed on 7 September 2020).

52. Giardinaggio Fagiolo-Phaseolus vulgaris-Orto-Fagiolo-Phaseolus Vulgaris—Orto. Available online: https://www.giardinaggio.it/orto/singoleorticole/fagiolo/fagiolo.asp (accessed on 7 September 2020).

53. Nicoletto, C.; Zanin, G.; Sambo, P.; Dalla Costa, L. Quality assessment of typical common bean genotypes cultivated in temperate climate conditions and different growth locations. *Sci. Hortic.* **2019**, *256*, 108599. [CrossRef]

54. Rocha-Guzmán, N.E.; González-Laredo, R.F.; Ibarra-Pérez, F.J.; Nava-Berúmen, C.A.; Gallegos-Infante, J.A. Effect of pressure cooking on the antioxidant activity of extracts from three common bean (*Phaseolus vulgaris* L.) cultivars. *Food Chem.* **2007**, *100*, 31–35. [CrossRef]

55. Boateng, J.; Verghese, M.; Walker, L.T.; Ogutu, S. Effect of processing on antioxidant contents in selected dry beans (*Phaseolus spp.* L.). *LWT Food Sci. Technol.* **2008**, *41*, 1541–1547. [CrossRef]

56. Barroga, C.F.; Laurena, A.C.; Mendoza, E.M.T. Polyphenols in Mung Bean (Vigna radiata (L.) Wilczek): Determination and Removal. *J. Agric. Food Chem.* **1985**, *33*, 1006–1009. [CrossRef]

57. Blair, M.W.; Astudillo, C.; Grusak, M.A.; Graham, R.; Beebe, S.E. Inheritance of seed iron and zinc concentrations in common bean (*Phaseolus vulgaris* L.). *Mol. Breed.* **2009**, *23*, 197–207. [CrossRef]

58. Amongi, W.; Mukankusi, C.; Sebuliba, S.; Mukamuhirwa, F. Iron and zinc grain concentrations diversity and agronomic performance of common bean germplasm collected from East Africa. *African J. Food Agric. Nutr. Dev.* **2018**, *18*, 13717–13742. [CrossRef]

59. Celmeli, T.; Sari, H.; Canci, H.; Sari, D.; Adak, A.; Eker, T.; Toker, C. The nutritional content of common bean (phaseolus vulgaris l.) landraces in comparison to modern varieties. *Agronomy* **2018**, *8*, 166. [CrossRef]

60. NIH (National Institutes of Health) Zinc—Consumer. Available online: https://ods.od.nih.gov/factsheets/Zinc-Consumer/ (accessed on 13 October 2020).

61. Paredes, M.; Becerra, V.; Tay, J. Inorganic Nutritional Composition of Common Bean (*Phaseolus vulgaris* L.) Genotypes Race Chile. *Chil. J. Agric. Res.* **2010**, *69*, 486–495. [CrossRef]

62. Bulyaba, R.; Winham, D.M.; Lenssen, A.W.; Moore, K.J.; Kelly, J.D.; Brick, M.A.; Wright, E.M.; Ogg, J.B. Genotype by Location Effects on Yield and Seed Nutrient Composition of Common Bean. *Agronomy* **2020**, *10*, 347. [CrossRef]

63. Kazai, P.; Noulas, C.; Khah, E.; Vlachostergios, D. Yield and seed quality parameters of common bean cultivars grown under water and heat stress field conditions. *AIMS Agric. Food* **2019**, *4*, 285–302. [CrossRef]

64. Davis, D.R.; Epp, M.D.; Riordan, H.D. Changes in USDA Food Composition Data for 43 Garden Crops, 1950 to 1999. *J. Am. Coll. Nutr.* **2004**, *23*, 669–682. [CrossRef] [PubMed]

65. Krüger, E.; Dietrich, H.; Hey, M.; Patz, C.D. Effects of cultivar, yield, berry weight, temperature and ripening stage on bioactive compounds of black currants. *J. Appl. Bot. Food Qual.* **2011**, *84*, 40–46.

66. Ribeiro, N.D.; Mambrin, R.B.; Storck, L.; Prigol, M.; Nogueira, C.W. Combined selection for grain yield, cooking quality and minerals in the common bean. *Rev. Ciência Agronômica* **2013**, *44*, 869–877. [CrossRef]

67. Ribeiro, N.D.; Jost, E.; Maziero, S.M.; Storck, L.; Rosa, D.P. Selection of common bean lines with high grain yield and high grain calcium and iron concentrations. *Rev. Ceres* **2014**, *61*, 77–83. [CrossRef]
68. Domingues, L.d.S.; Ribeiro, N.D.; Andriolo, J.L.; Possobom, M.T.D.F.; Zemolin, A.E.M. Growth, grain yield and calcium, potassium and magnesium accumulation in common bean plants as related to calcium nutrition. *Acta Sci. Agron.* **2016**, *38*, 207–217. [CrossRef]

 *horticulturae*

Article

Gibberellic Acid Induced Changes on Growth, Yield, Superoxide Dismutase, Catalase and Peroxidase in Fruits of Bitter Gourd (*Momordica charantia* L.)

Mazhar Abbas [1,*], Faisal Imran [1], Rashid Iqbal Khan [1], Muhammad Zafar-ul-Hye [2], Tariq Rafique [3], Muhammad Jameel Khan [4], Süleyman Taban [5], Subhan Danish [2] and Rahul Datta [6,*]

[1] Institute of Horticultural Sciences, University of Agriculture Faisalabad, Punjab 38000, Pakistan; imranuaf555@gmail.com (F.I.); rashid.khan3535@yahoo.com (R.I.K.)
[2] Department of Soil Science, Faculty of Agricultural Sciences and Technology, Bahauddin Zakariya University Multan, Punjab 60800, Pakistan; zafarulhye@bzu.edu.pk (M.Z.-u.-H.); sd96850@gmail.com (S.D.)
[3] PGRP, Bio-resource Conservation Institute, NARC, Islamabad 44000, Pakistan; tariqabp@gmail.com
[4] Department of Soil Science and Environmental Science, Gomal University, Dera Ismail Khan, Khyber Pakhtunkhwa 29050, Pakistan; jamil@gu.edu.pk
[5] Department of Soil Science and Plant Nutrition, Faculty of Agriculture, Ankara University, Dışkapı-Ankara 06110, Turkey; taban@agri.ankara.edu.tr
[6] Department of Geology and Pedology, Mendel University in Brno, Zemedelska1, Brno 61300, Czech Republic
* Correspondence: rmazhar@hotmail.com (M.A.); Tel.: +92-3216685249 (M.A.); rahulmedcure@gmail.com (R.D.); Tel.: +42-0773990283 (R.D.)

Received: 16 September 2020; Accepted: 12 October 2020; Published: 29 October 2020

Abstract: Bitter gourd is one of the important cucurbits and highly liked among both farmers and consumers due to its high net return and nutritional value. However, being monoecious, it exhibits substantial variation in flower bearing pattern. Plant growth regulators (PGRs) are known to influence crop phenology while gibberellic acid (GA$_3$) is one of the most prominent PGRs that influence cucurbits phenology. Therefore, a field trial was conducted at University of Agriculture Faisalabad to evaluate the impact of a commercial product of gibberellic acid (GA$_3$) on growth, yield and quality attributes of two bitter gourd (*Momordica charantia* L.) cultivars. We used five different concentrations (0.4 g, 0.6 g, 0.8 g, 1.0 g, and 1.2 g per litre) of commercial GA$_3$ product (Gibberex, 10% Gibberellic acid). Results showed that a higher concentration of gibberex (1.0 and 1.20 g L^{-1} water) enhanced the petiole length, intermodal length, and yield of bitter gourd cultivars over control in Golu hybrid and Faisalabad Long. A significant decrease in the enzyme superoxidase dismutase, peroxidase, and catalase activities were observed with an increasing concentration of gibberex (1.0 and 1.20 gL^{-1} water) as compared to control. These results indicate that the exogenous application of gibberex at a higher concentration (1.2 g L^{-1}) has a dual action in bitter gourd plant: i) it enhances the plant growth and yield, and ii) it also influenced the antioxidant enzyme activities in fruits. These findings may have a meaningful, practical use for farmers involved in agriculture and horticulture.

Keywords: gibberex; *Momordica charantia* L; dismutase; peroxidase; catalase; vegetative growth

---

1. Introduction

Bitter gourd (*Momordica charantia* L.), well-known as bitter melon, balsam pear, and bitter cucumber [1] exhibits high nutritional value in terms of enriched mineral, vitamins, and macro-nutrient contents [2]. Despite providing essential elements it also possesses antioxidant, antiviral and antimicrobial pressure in addition to lowering blood sugar makes it favourite among consumers [3]. Meanwhile, farmers prefer its cultivation due to shorter growing period, early maturity, incremented

2.5. Statistical Analysis

Randomized Complete Block Design (RCBD) was used to evaluate the impact of five treatments and control having three replications with three plants per replication to conduct this experiment. Data were analyzed with the help of two way ANOVA (analysis of variance) and means were compared by applying LSD test at 5% level of significance [40].

3. Results and Discussion

3.1. Petiole and Internodal Length

Results indicated that gibberex application as treatment significantly ($p \leq 0.05$) increased the internodal length (IL) and petiole length (PL) of Golu hybrid and Faisalabad Long (Figure 1A,B). Interaction of different application rates of gibberex and varieties of bitter gourd was ordinal for petiole length (Figure 2A). A positive non-significant correlation was observed between petiole length of Golu hybrid and Faisalabad Long with different application rates of gibberex (Figure 2A). Application of treatment 1.0 gL^{-1} gibberex remained significant as compared to control in Golu hybrid for petiole length. In Faisalabad Long, treatments 1.2 and 0.6 gL^{-1} gibberex showed a significant positive response for improvement in petiole length. Maximum petiole length (9.33 and 7.33 cm) was recorded in Golu hybrid and Faisalabad Long at 1.0 and 1.2 gL^{-1} gibberex respectively.

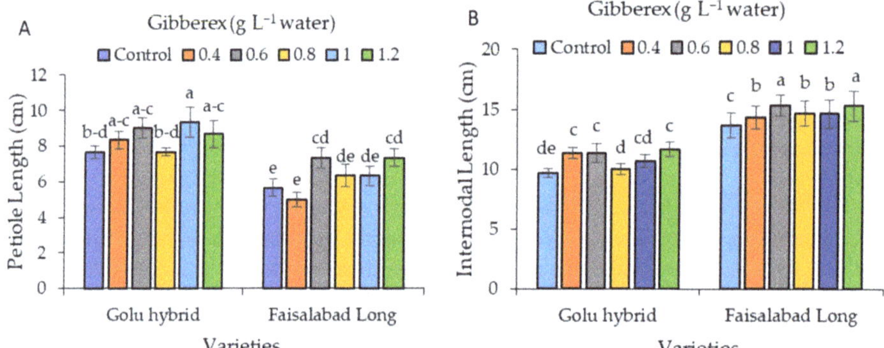

Figure 1. Effect of concentrations of Gibberex on petiole length (**A**) and Internodal length (**B**) of two bitter gourd cultivars. Vertical Bars indicate standard error of means (n = 9). Means sharing different letters are significantly different from each other. * Significant ($p \leq 0.05$), Highly significant ($p \leq 0.01$) while results depicts that for petiole length: Varieties = **, GA = *, Varieties × GA = NS; for Internodal Length: Varieties = **, GA = *, Varieties × GA = NS.

For intermodal length, 1.2 gL^{-1} gibberex differed significantly over control in Golu hybrid and Faisalabad Long. Interaction of different application rates of gibberex and varieties of bitter gourd was disordinal for internodal length (Figure 2B). A positive non-significant correlation was observed between intermodal lengths of Golu hybrid with different application rates of gibberex (Figure 9A). However, intermodal length of Faisalabad Long showed positive significant correlation with different application rates of gibberex (Figure 9B). All application rates of gibberex remained significant over control in Faisalabad Long for internodal length. For Golu hybrid, treatments 0.4 and 0.6 gL^{-1} gibberex also remained equally significant over control for internodal length. Maximum internodal length (15.33 and 11.66 cm) was recorded in Faisalabad Long and Golu hybrid at 1.2 gL^{-1} gibberex respectively. Internodal elongation is based on the ability of plants to perform cell division and cell elongation [41] that varies from genotype to genotype. In contrast, this increment could also be due to the application of gibberex stimulated cell division and cell elongation [42] leading to stem elongation via stretching internodal distance [43]. These findings are comparable with outcomes of Bostrack and

Struckmeyer [44] that observed elongation of pith parenchyma cells leading to elongation of internodal length under GA_3 application. Similar findings were reported by Ros et al. [45] and Mishra et al. [13] that GA_3 application enhanced the internodal distance and petiole length in pea and bitter gourd.

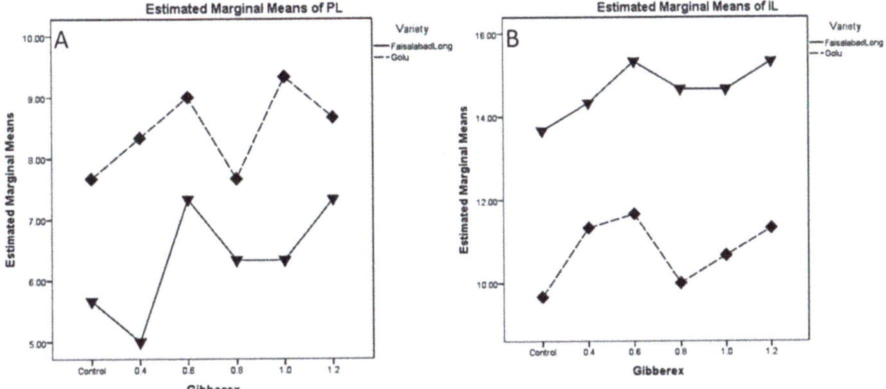

Figure 2. Interaction graph of different application rates of gibberex and bitter gourd varieties for petiole and intermodal length. Figure (**A**) represents petiole length while (**B**) represents the internodal length.

3.2. Fruit Number and Weight

Effect of different application rates of Gibberex was significant ($p \leq 0.05$) on number of fruits (NF) (Figure 3A) and fruit weight (FW) (Figure 3B) of Golu hybrid and Faisalabad Long. Interaction of different application rates of Gibberex and varieties of bitter gourd was disordinal for number of fruits (Figure 4A). A positive significant correlation was observed between Golu hybrid and Faisalabad number of fruits with different application rates of gibberex (Figure 9A,B). It was noted that application of treatment 1.2 gL^{-1}. Gibberex was significantly different as compared to other application rates of Gibberex for improvement in number of fruits of Golu hybrid and Faisalabad Long. However, number of fruits were significantly higher in Golu hybrid over Faisalabad Long. For the increase in fruit weight of Golu hybrid and Faisalabad Long, higher application rate 1.2 gL^{-1} Gibberex as treatment differed significantly from control. Interaction of different application rates of Gibberex and varieties of bitter gourd was significant ordinal for fruit weight (Figure 4B). A positive significant correlation was observed between Golu hybrid and Faisalabad fruit weight with different application rates of gibberex (Figure 9A,B). Maximum increase of 17.8 and 31.4% in number of fruits and 26.4 and 36.1% in fruit weight were noted where 1.2 gL^{-1} gibberex in Golu hybrid and Faisalabad Long respectively. Balance uptake of nutrients by application of Gibberex played an imperative role in the improvement of number of fruits and fruit weight in bitter gourd varieties [43]. Application of GA_3 enhanced the synthesis of chlorophyll which led to enhanced production of proteins [46] whereas varietal genetic potential to produce a low or high number of fruits can also affect fruit number per plant. Moreover, Sadia et al. [47] concluded another factor that GA_3 enhances the initial growth of ovaries along with reducing the abscission layer that leads to a reduction premature flower and fruit drop.

3.3. Yield per Plant

Results indicated a significant difference ($p \leq 0.05$) among the bitter gourd cultivars for yield per plant (YP) (Figure 5A). A significant ordinal interaction was observed among different varieties of bitter gourd and different application rates of gibberex (Figure 5B). A positive significant correlation was observed between Golu hybrid and Faisalabad Long regarding yield per plant with different application rates of gibberex (Figure 9A,B). It was observed that the application of 1.0 and 1.2 gL^{-1} gibberex proved to be significantly best application rate over control for the improvement in the yield per plant in Golu hybrid over Faisalabad Long respectively. Maximum increases of 41% and 88% in yield per plant were noted where 1.0 and 1.2 gL^{-1} gibberex treatments were applied in Golu hybrid and Faisalabad Long respectively. Role of GA$_3$ had been in focus in various plant species for increasing biomass and yields [48,49] as its exogenous application has proved to promote source and sink relation with an additional increment of dry matter accumulation. However, a high increment of yield in Faisalabad long following gibberex application could be ascribed to both improved genetic traits and GA$_3$ mediated enhanced nutritional uptake subsequently promoting the leaves photosynthetic capacity [50]. Moreover, GAs had also been set up to facilitate the integration of source to the developing reproductive plant parts followed by enhanced sink strength due to improved photosynthate translocation [51]. Further, GA$_3$ enhances the transformation of sucrose to glucose, thus improving extracellular invertase [52] that leads to phloem delivering carbohydrates to sink organs, thus ultimately enhancing plant yield.

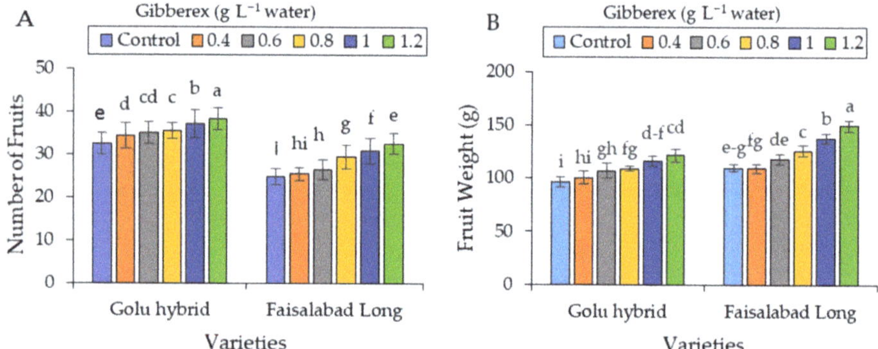

Figure 3. Effect of concentrations of Gibberex on number of fruits (**A**) and fruit weight (**B**) of two bitter gourd cultivars. Vertical Bars indicate standard error of means (n = 9). Means sharing different letters are significantly different from each other. * Significant ($p \leq 0.05$), ** Highly significant ($p \leq 0.01$). while statistical analysis depicts that results for number of fruits: Varieties = **, GA = **, Varieties $\times$ GA = **; for Fruit weight: Varieties = **, GA = **, Varieties $\times$ GA = *.

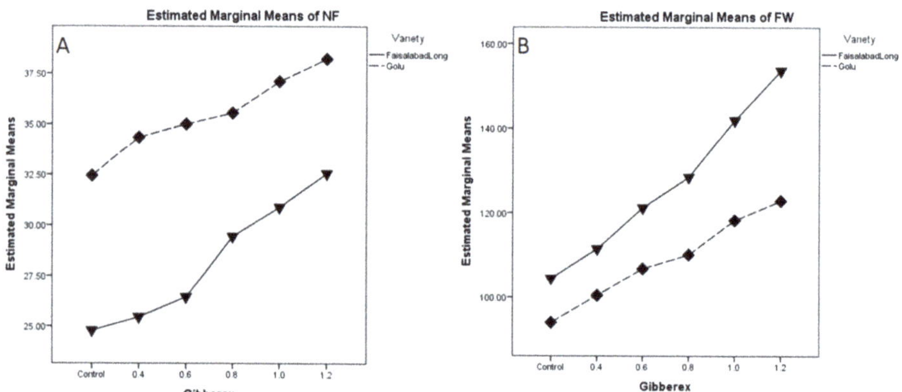

Figure 4. Interaction graph of different application rates of gibberex and bitter gourd varieties for number of fruits and fruit weight. Figure (**A**) represents number of fruit while (**B**) represents fruit weight.

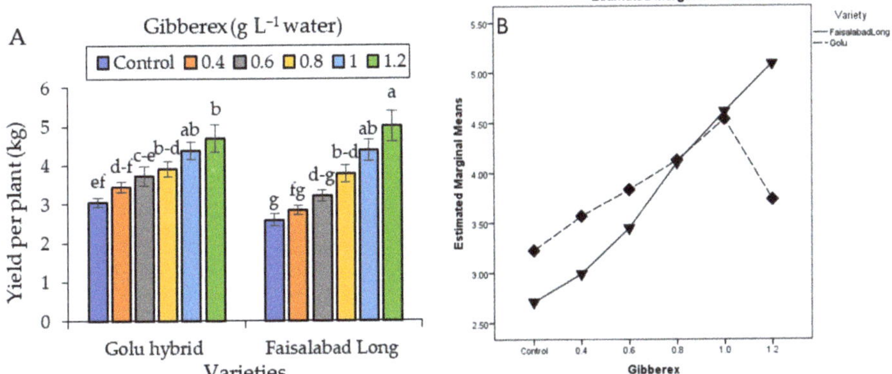

Figure 5. Effect of concentration of Gibberex on yield per plant of two bitter gourd cultivars. Figure (**A**) represents yield per plant while (**B**) represents the interaction graph of different application rates of gibberex and bitter gourd varieties for yield per plant. Vertical Bars indicate standard error of means (n = 9). Means sharing different letters are significantly different from each other. * Significant ($p \leq 0.05$), ** Highly significant ($p \leq 0.01$) while results depicted that for yield per plant: Varieties = NS, GA = **, Varieties × GA = **.

3.4. SOD, POD and CAT

Effect of different application rates of gibberex significantly ($p \leq 0.05$) affect the SOD (Figure 6A), POD (Figure 7A) and CAT (Figure 8A) in bitter gourd varieties. Interaction between different application rates of gibberex and bitter gourd varieties was significant ordinal for SOD (Figure 6B), POD (Figure 7B) and CAT (Figure 8B). A negative significant correlation was observed between Golu hybrid and Faisalabad SOD, POD and CAT with different application rates of gibberex (Figure 9A,B). Increasing levels of gibberex significantly decreased the SOD, POD and CAT in Golu hybrid and Faisalabad Long. Significant reductions in SOD (79% and 72%), POD (62% and 61%), and CAT (55% and 51%) were noted where treatment 1.2 gL⁻¹ gibberex was applied in Golu hybrid and Faisalabad Long respectively. These findings are comparable with the outcomes of Rosenwasser et al. [53]. They also reported that exogenously applied GA_3 reduced the production of reactive oxygen species and thus leads to improvement in the activity of SOD in *Catharanthus roseus*. Previous studies have concluded that GA_3 application can enhance the antioxidant activity of plants by inducing phenolic production [54] reducing lipid peroxidation with increasing proline and glycine-betaine biosynthesis [55]. Previous studies have reported that plant growth regulators alone or in combination are involved in regulating the transcription of genes [56] along with mRNA [57] that ultimately, in turn, improve the synthesis of a specific hormone to make proteins.

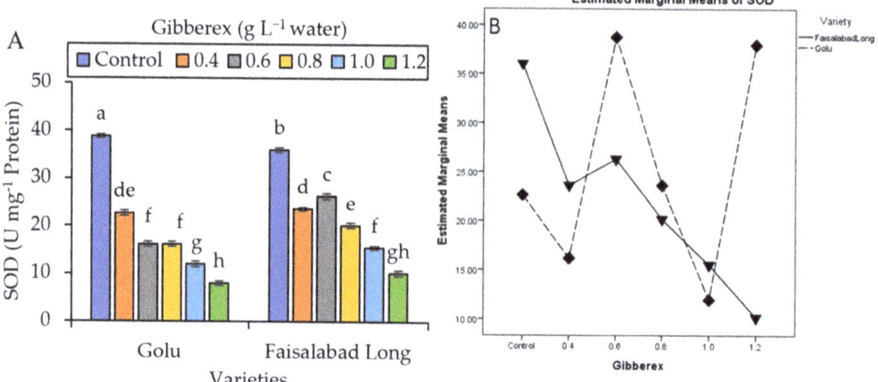

Figure 6. Effect of different concentrations of Gibberex on SOD of two bitter gourd cultivars. Figure (**A**) represents SOD while (**B**) represents the interaction graph of different application rates of gibberex and bitter gourd varieties for SOD. Vertical Bars indicate standard error of means (n = 9). Means sharing different letters are significantly different from each other. * Significant ($p \leq 0.05$), ** Highly significant ($p \leq 0.01$) while results depicted that for SOD activity: Varieties = **, GA = **, Varieties × GA = **.

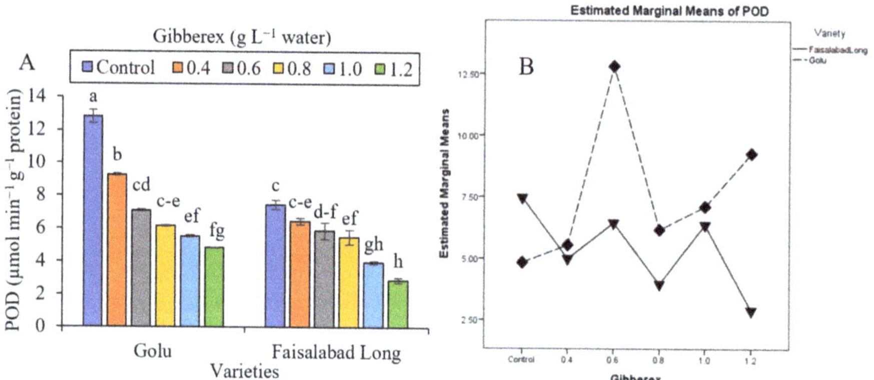

Figure 7. Effect of different concentrations of Gibberex on POD of two bitter gourd cultivars. Figure (**A**) represents POD while (**B**) represents the interaction graph of different application rates of gibberex and bitter gourd varieties for POD. Vertical Bars indicate standard error of means (n = 9). Means sharing different letters are significantly different from each other. * Significant ($p \leq 0.05$), ** Highly significant ($p \leq 0.01$) while results depicted that for POD activity: Varieties = **, GA = **, Varieties × GA = **.

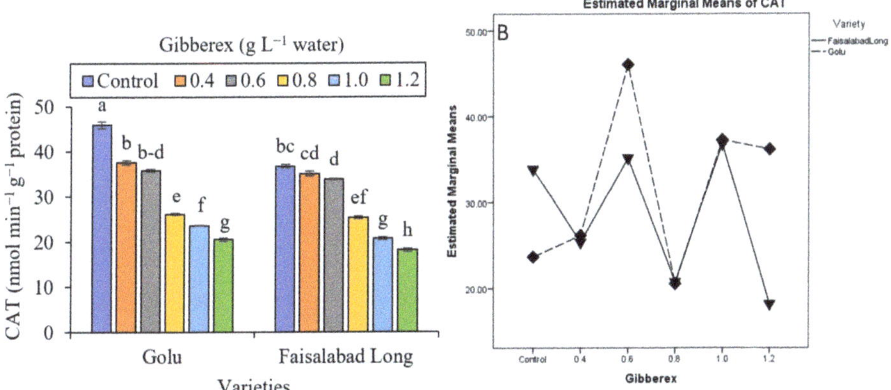

Figure 8. Effect of different concentrations of Gibberex on CAT of two bitter gourd cultivars. Figure (**A**) represents CAT while (**B**) represents the interaction graph of different application rates of gibberex and bitter gourd varieties for CAT. Vertical Bars indicate standard error of means (n = 9). Means sharing different letters are significantly different from each other. * Significant ($p \leq 0.05$), ** Highly significant ($p \leq 0.01$) while results depicted that for yield per plant: Varieties = **, GA = **, Varieties × GA = **.

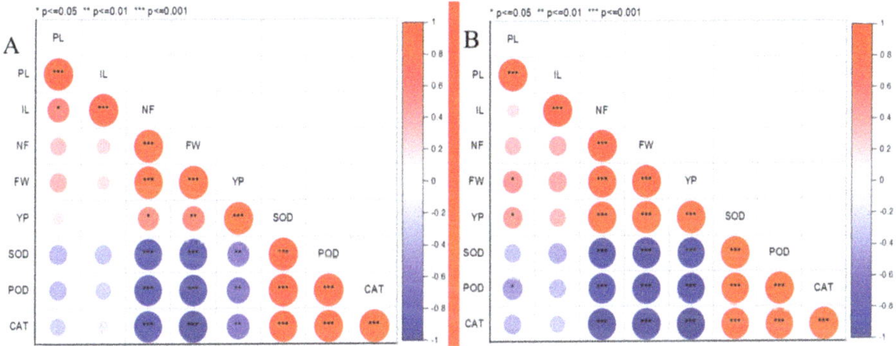

Figure 9. Pearson correlation between different application rates of gibberex and bitter gourd growth, yield and antioxidants. Figure (**A**) represents Golu hybrid and (**B**) represents Faisalabad Long. Red circles showed a positive correlation with attributes. Blue circles showed a negative correlation with attributes. * Significant ($p \leq 0.05$), ** Highly significant ($p \leq 0.01$) while results depicted that for yield per plant: Varieties = **, GA = **, Varieties × GA = **.

4. Conclusions

It is concluded that 1.0 and 1.2 gL^{-1} gibberex application lead to significant enhancement of petiole and intermodal length in bitter gourd. In particular, it enhances the number of fruits with heavier fruit weight hence increasing yield and giving good returns to the growers. Application of 1.0 and 1.2 gL^{-1} gibberex is also efficacious in decreasing SOD, POD, and CAT in Golu hybrid and Faisalabad Long. However, further research is suggested with molecular analysis to have a detailed insight into gibberellic acid induced changes responsible for the enhancement of growth and yield in Golu hybrid and Faisalabad Long.

Author Contributions: Conceptualization, M.A.; methodology, M.J.K., F.I. and T.F.; formal analysis, M.J.K., M.A. and R.I.K.; investigation S.D., M.A., and R.D.; writing—original draft preparation, S.T., M.Z.-u.-H., R.I.K., T.R., and S.D.; writing—review and editing, M.A.; supervision, M.A. All authors have read and agreed to the published version of the manuscript.

Funding: This research received no external funding.

Conflicts of Interest: The authors declare no conflict of interest.

References

1. Ahmad, N.; Hasan, N.; Ahmad, Z.; Zishan, M.; Zohrameena, S. Momordica charantia: For traditional uses and pharmacological actions. *J. Drug Deliv. Ther.* **2016**, *6*. [CrossRef]
2. Bortolotti, M.; Mercatelli, D.; Polito, L. Momordica charantia, a nutraceutical approach for inflammatory related diseases. *Front. Pharmacol.* **2019**, *10*, 486. [CrossRef]
3. Salehi, B.; Ata, A.; Anil Kumar, V.N.; Sharopov, F.; Ramírez-Alarcón, K.; Ruiz-Ortega, A.; Abdulmajid Ayatollahi, S.; Tsouh Fokou, P.V.; Kobarfard, F.; Amiruddin Zakaria, Z.; et al. Antidiabetic potential of medicinal plants and their active components. *Biomolecules* **2019**, *9*, 551. [CrossRef]
4. Adil, S.A.; Chattha, M.W.A.; Bakhsh, K.; Hassan, S. Profitability analysis of summer vegetables by farm size. *Pak. J. Agri. Sci* **2007**, *44*, 184–188.
5. Mila, F.A.; Rahman, M.S.; Nahar, N.; Debnath, D.; Shahjada, M.K.H. Profitability of bitter gourd production in some areas of Narsingdi District. *J. Sylhet Agric. Univ.* **2015**, *2*, 143–147.
6. Behera, T.K.; Behera, S.; Bharathi, L.K.; John, K.J.; Simon, P.W.; Staub, J.E. Bitter gourd: Botany, horticulture, breeding. *Hortic. Rev.* **2010**, *37*, 101–141.
7. Rasco, A.O.; Castillo, P.S. Flowering patterns and vine pruning effects in bittergourd (Momordica charantia L) varieties "Sta. Rita" and "Makiling.". *Philipp. Agric.* **1990**, *73*, 311–322.
8. Dey, S.S.; Batters, T.K.; Pal, A.; Munshi, A.D. Correlation and path coefficient analysis in bitter gourd. *Veg. Sci.* **2005**, *32*, 173–176.
9. Li, S.F.; Zhang, G.J.; Yuan, J.H.; Deng, C.L.; Gao, W.J. Repetitive sequences and epigenetic modification: Inseparable partners play important roles in the evolution of plant sex chromosomes. *Planta* **2016**, *243*, 1083–1095. [CrossRef]
10. Lai, Y.-S.; Shen, D.; Zhang, W.; Zhang, X.; Qiu, Y.; Wang, H.; Dou, X.; Li, S.; Wu, Y.; Song, J.; et al. Temperature and photoperiod changes affect cucumber sex expression by different epigenetic regulations. *BMC Plant Biol.* **2018**, *18*, 268. [CrossRef]
11. Girek, Z.; Prodanovic, S.; Zdravkovic, J.; Zivanovic, T.; Ugrinovic, M.; Zdravkovic, M. The effect of growth regulators on sex expression in melon (Cucumis melo L.). *Crop. Breed. Appl. Biotechnol.* **2013**, *13*, 165–171. [CrossRef]
12. Khan, N.; Bano, A.; Babar, M.D.A. Metabolic and physiological changes induced by plant growth regulators and plant growth promoting rhizobacteria and their impact on drought tolerance in Cicer arietinum L. *PLoS ONE* **2019**, *14*, e0213040. [CrossRef] [PubMed]
13. Mishra, S.; Behera, T.K.; Munshi, A.D.; Bharadwaj, C.; Rao, A.R. Inheritance of gynoecism and genetics of yield and yield contributing traits through generation mean analysis in bitter gourd. *Indian J. Hortic.* **2015**, *72*, 218–222. [CrossRef]
14. Pawełkowicz, M.; Pryszcz, L.; Skarzyńska, A.; Wóycicki, R.K.; Posyniak, K.; Rymuszka, J.; Przybecki, Z.; Pląder, W. Comparative transcriptome analysis reveals new molecular pathways for cucumber genes related to sex determination. *Plant Reprod.* **2019**, *32*, 193–216. [CrossRef] [PubMed]
15. Zhang, Y.; Zhao, G.; Li, Y.; Mo, N.; Zhang, J.; Liang, Y. Transcriptomic analysis implies that GA regulates sex expression via ethylene-dependent and ethylene-independent pathways in cucumber (Cucumis sativus L.). *Front. Plant Sci.* **2017**, *8*, 10. [CrossRef]
16. Molaei, A.; Lakzian, A.; Haghnia, G.; Astaraei, A.; Rasouli-Sadaghiani, M.; Ceccherini, M.T.; Datta, R. Assessment of some cultural experimental methods to study the effects of antibiotics on microbial activities in a soil: An incubation study. *PLoS ONE* **2017**, *12*, e0180663. [CrossRef]
17. Molaei, A.; Lakzian, A.; Datta, R.; Haghnia, G.; Astaraei, A.; Rasouli-Sadaghiani, M.; Ceccherini, M.T. Impact of chlortetracycline and sulfapyridine antibiotics on soil enzyme activities. *Int. Agrophys.* **2017**, *31*, 499–505. [CrossRef]

18. Meena, R.S.; Kumar, S.; Datta, R.; Lal, R.; Vijayakumar, V.; Brtnicky, M.; Sharma, M.P.; Yadav, G.S.; Jhariya, M.K.; Jangir, C.K. Impact of agrochemicals on soil microbiota and management: A review. *Land* **2020**, *9*, 34. [CrossRef]

19. Brtnicky, M.; Holatko, J.; Pecina, V.; Kintl, A.; Latal, O.; Vyhnanek, T.; Prichystalova, J.; Datta, R. Long-term effects of biochar-based organic amendments on soil microbial parameters. *Agronomy* **2019**, *9*, 747. [CrossRef]

20. Aya, K.; Ueguchi-Tanaka, M.; Kondo, M.; Hamada, K.; Yano, K.; Nishimura, M.; Matsuoka, M. Gibberellin modulates anther development in rice via the transcriptional regulation of GAMYB. *Plant Cell* **2009**, *21*, 1453–1472. [CrossRef]

21. Song, S.; Qi, T.; Huang, H.; Xie, D. Regulation of stamen development by coordinated actions of jasmonate, auxin, and gibberellin in arabidopsis. *Mol. Plant* **2013**, *6*, 1065–1073. [CrossRef]

22. Benková, E. Plant hormones in interactions with the environment. *Plant Mol. Biol.* **2016**, *91*, 597. [CrossRef] [PubMed]

23. Crow, J.R.; Thomson, R.J.; Mander, L.N. Synthesis and confirmation of structure for the gibberellin GA 131 (18-hydroxy-GA4). *Org. Biomol. Chem.* **2006**, *4*, 2532–2544. [CrossRef] [PubMed]

24. Gupta, R.; Chakrabarty, S.K. Gibberellic acid in plant: Still a mystery unresolved. *Plant Signal. Behav.* **2013**, *8*, e25504. [CrossRef] [PubMed]

25. Anwar, A.; Yan, Y.; Liu, Y.; Li, Y.; Yu, X. 5-Aminolevulinic acid improves nutrient uptake and endogenous hormone accumulation, enhancing low-temperature stress tolerance in cucumbers. *Int. J. Mol. Sci.* **2018**, *19*, 3379. [CrossRef] [PubMed]

26. Hamdy, A. Effect of GA3 and NAA on growth, yield and fruit quality of Washington navel orange. *Egypt. J. Hortic.* **2017**, *44*, 33–43. [CrossRef]

27. Shah, S.-U.-S.; Khan, A.; Khan, M.A.; Farooq, K.; Riaz Chattha, M.; Javed, M.A.; Gurmani, Z.A.; Hussain, A.; Iftikhar, M. Effects of gibberellic acid on growth, yield and quality of grape cv. *Perlet* **2015**, *12*, 499–503.

28. Tantasawat, P.A.; Sorntip, A.; Pornbungkerd, P. Effects of exogenous application of plant growth regulators on growth, yield, and In Vitro gynogenesis in cucumber. *HortScience* **2015**, *50*, 374–382. [CrossRef]

29. Ghani, M.A.; Amjad, M.; Iqbal, Q.; Nawaz, A.; Ahmad, T.; Hafeez, O.B.A.; Abbas, M. Efficacy of plant growth regulators on sex expression, earliness and yield components in bitter gourd. *Pakistan J. Life Soc. Sci.* **2013**, *11*, 218–224.

30. Hossain, D.; Abdul Karim, M.; Rahman Pramanik, M.H.; Syedur Rahman, A.A.M. Effect of ld (GA3) on flowering and fruit development of bitter gourd (Momordica charantia L.). *Int. J. Bot.* **2006**, *2*, 329–332. [CrossRef]

31. Nagmani, S. Studies on Crop Growth and Sex Expression in Relation to Hybrid Seed Production of Bitter Gourd (Momordica charinata L.). Master's Thesis, Faculty of Post Graduate Schol, IARI, New Delhi, India, 2011.

32. Marfo, T.D.; Vranová, V.; Ekielski, A. Ecotone dynamics and stability from soil perspective: Forest-agriculture land transition. *Agriculture* **2019**, *9*, 228. [CrossRef]

33. Marfo, T.D.; Datta, R.; Pathan, S.I.; Vranová, V. Ecotone dynamics and stability from soil scientific point of view. *Diversity* **2019**, *11*, 53. [CrossRef]

34. Yadav, G.S.; Datta, R.; Pathan, S.I.; Lal, R.; Meena, R.S.; Babu, S.; Das, A.; Bhowmik, S.N.; Datta, M.; Saha, P.; et al. Effects of conservation tillage and nutrient management practices on soil fertility and productivity of rice (Oryza sativa L.)-rice system in North eastern region of India. *Sustainability* **2017**, *9*, 1816. [CrossRef]

35. Gocmen, M.; Aydin, E.; Simsek, I.; Sari, N.; Solmaz, I.; Gokseven, A. Characterization of some agronomic traits and β-carotene contents of orange fleshed altinbas melon dihaploid lines. *Ekin J. Crop Breed. Genet.* **2017**, *3*, 12–18.

36. Yagoub, S.O.; Ahmed, W.M.A.; Mariod, A.A. Show more effect of urea, NPK and compost on growth and yield of soybean (Glycine max L), in Semi-Arid Region of Sudan. *Int. Sch. Res. Not.* **2012**, *2012*, 678124.

37. El-Kholy, E.; Hafez, H. Response of snake cucumber (Cucumis melo var. pubescens) to some plant regulators. *J. Agric. Sci.* **1982**, *99*, 587–590. [CrossRef]

38. Štajner, D.; Popović, B. Comparative study of antioxidant capacity in organs of different Allium species. *Open Life Sci.* **2009**, *4*, 224–228. [CrossRef]

39. Chance, B.; Maehly, A.C. Assay of catalases and peroxidases. In *Methods of Biochemical Analysis*; John Wiley & Sons, Inc.: Hoboken, NJ, USA, 1954; Volume 1.

40. Steel, R.G.; Torrie, J.H.; Dickey, D.A. *Principles and Procedures of Statistics: A Biometrical Approach*, 3rd ed.; McGraw Hill Book International Co.: Singapore, 1997.

41. Anandan, A.; Rajiv, G.; Ramarao, A.; Prakash, M. Internode elongation pattern and differential response of rice genotypes to varying levels of flood water. *Funct. Plant Biol.* **2012**, *39*, 137–145. [CrossRef]

42. Amooaghaie, R. The effect mechanism of moist-chilling and GA3 on seed germination and subsequent seedling growth of ferula ovina boiss. *Open Plant Sci. J.* **2009**, *3*, 22–28. [CrossRef]

43. Ullah, R.; Sajid, M.; Ahmad, H.; Luqman, M.; Razaq, M.; Nabi, G.; Fahad, S.; Rab, A. Association of gibberellic acid (GA3) with fruit set and fruit drop of sweet orange. *J. Biol.* **2014**, *4*, 54–59.

44. Bostrack, J.M.; Struckmeyer, B.E. Effect of gibberellic acid on the growth and anatomy of coleus blumei, antirrhinum majus and salvia splendens. *N. Phytol.* **1967**, *66*, 539–544. [CrossRef]

45. Ross, J.J.; O'Neill, D.P.; Rathbone, D.A. Auxin-gibberellin interactions in pea: Integrating the old with the new. *J. Plant Growth Regul.* **2003**, *22*, 99–108. [CrossRef]

46. Ramteke, V.; Paithankar, D.H.; Baghel, M.M.; Kurrey, V.K. Impact of GA3 and propagation media on growth rate and leaf chlorophyll content of papaya seedlings. *Res. J. Agric. Sci.* **2016**, *7*, 169–171.

47. Sadia, S.; Rab, A.; Shah, S.H.A.; Ullah, I.; Bibi, F.; Zeb, I. Influence of De-blossoming and GA3 application on fruit drop and growth of winter guava. *Sarhad J. Agric.* **2017**, *33*, 526–531. [CrossRef]

48. Pan, S.; Rasul, F.; Li, W.; Tian, H.; Mo, Z.; Duan, M.; Tang, X. Roles of plant growth regulators on yield, grain qualities and antioxidant enzyme activities in super hybrid rice (Oryza sativa L.). *Rice* **2013**, *6*, 9. [CrossRef]

49. Whitehead, D.; Edwards, G.R. Assessment of the application of gibberellins to increase productivity and reduce nitrous oxide emissions in grazed grassland. *Agric. Ecosyst. Environ.* **2015**, *207*, 40–50. [CrossRef]

50. Eid, R.A.; Abou-Leila, B.H. Response of croton plants to gibberellic acid, benzyl adenine and ascorbic acid application. *World J. Agric. Sci.* **2006**, *2*, 174–179.

51. Khan, N.A.; Singh, S.; Nazar, R.; Lone, P.M. The source sink relationship in mustard. *Asian Australas. J. Plant Sci. Biotechnol.* **2007**, *1*, 10–18.

52. Zhang, C.; Tanabe, K.; Tamura, F.; Itai, A.; Yoshida, M. Roles of gibberellins in increasing sink demand in Japanese pear fruit during rapid fruit growth. *Plant Growth Regul.* **2007**, *52*, 161–172. [CrossRef]

53. Rosenwasser, S.; Belausov, E.; Riov, J.; Holdengreber, V.; Friedman, H. Gibberellic acid (GA3) inhibits ROS increase in chloroplasts during dark-induced senescence of pelargonium cuttings. *J. Plant Growth Regul.* **2010**, *29*, 375–384. [CrossRef]

54. Liang, Z.; Ma, Y.; Xu, T.; Cui, B.; Liu, Y.; Guo, Z.; Yang, D. Effects of abscisic acid, gibberellin, ethylene and their interactions on production of phenolic acids in salvia miltiorrhiza bunge hairy roots. *PLoS ONE* **2013**, *8*, e72806. [CrossRef]

55. Ahmad, P. Growth and antioxidant responses in mustard (Brassica junceaL.) plants subjected to combined effect of gibberellic acid and salinity. *Arch. Agron. Soil Sci.* **2010**, *56*, 575–588. [CrossRef]

56. Hoad, G.V.; Lenton, J.R.; Jackson, M.B. *Hormone Action in Plant Development–A Critical Appraisal*; Butterworth-Heinemann: Oxford, UK, 1987.

57. Wu, L.L.; Mitchell, J.P.; Cohn, N.S.; Kaufman, P.B. Gibberellin (GA3) enhances cell wall invertase activity and mRNA levels in elongating dwarf pea (Pisum sativum) shoots. *Int. J. Plant Sci.* **1993**, *154*, 280–289. [CrossRef]

Publisher's Note: MDPI stays neutral with regard to jurisdictional claims in published maps and institutional affiliations.

 © 2020 by the authors. Licensee MDPI, Basel, Switzerland. This article is an open access article distributed under the terms and conditions of the Creative Commons Attribution (CC BY) license (http://creativecommons.org/licenses/by/4.0/).

 *horticulturae*

Article

Effect of Seaweed Extract on Productivity and Quality Attributes of Four Onion Cultivars

Mazhar Abbas [1], Jahanzeb Anwar [1], Muhammad Zafar-ul-Hye [2], Rashid Iqbal Khan [1], Muhammad Saleem [3], Ashfaq Ahmad Rahi [4], Subhan Danish [2] and Rahul Datta [5,*]

1   Institute of Horticultural Sciences, Faculty of Agriculture, University of Agriculture, Faisalabad 38000, Pakistan; rmazhar@hotmail.com (M.A.); muhammadjahanzebanwar@gmail.com (J.A.); rashid.khan3535@yahoo.com (R.I.K.)
2   Department of Soil Science, Faculty of Agricultural Sciences & Technology, Bahauddin Zakariya University, Multan 60800, Punjab, Pakistan; zafarulhyegondal@yahoo.com (M.Z.-u.-H.); sd96850@gmail.com (S.D.)
3   Soil and Water Testing Laboratory, Vehari 61100, Punjab, Pakistan; saleembhatti1234@gmail.com
4   Pesticide Quality Control Laboratory, Multan 60000, Punjab, Pakistan; rahisenior2005@gmail.com
5   Department of Agrochemistry, Soil Science, Microbiology and Plant Nutrition, Faculty of AgriSciences, Mendel University in Brno, Zemedelska 1, 61300 Brno, Czech Republic
*   Correspondence: rahulmedcure@gmail.com; Tel.: +420-773990283

Received: 27 March 2020; Accepted: 2 May 2020; Published: 8 May 2020

Abstract: The excessive use of chemicals and inorganic fertilizers by farmers to increase crop yield is detrimental to the environment and human health. Application of biostimulants such as seaweed extract (SWE) in agriculture could be an effective and eco-friendly alternative to inorganic fertilizers. Biostimulants are natural organic degradable substances. Their application serves as a source of nutrition for crops, possibly improving growth and productivity when applied in combination with the fertilizers. The current study was conducted to evaluate the vegetative growth, reproductive behavior and quality attributes of four onion cultivars, 'Lambada', 'Red Bone', 'Nasarpuri', and 'Phulkara', in response to different concentrations of commercial SWE. Four levels of SWE extract were used, 0% (control), 0.5%, 1%, 2%, and 3%, which were applied as a foliar spray to each cultivar. The application of 0.5% SWE caused a significant increase in total soluble solids, mineral content (N, P, and K), bulb weight and yield. Application at 3% SWE increased ascorbic acid as compared to control. The cultivars responded in different ways regarding bulb dry weight and bulb and neck diameter. Among all cultivars, 'Lambada' showed the maximum bulb dry matter, 'Phulkara' showed enhanced neck diameter whereas 'Red Bone' showed maximum leaf length. It is concluded that 0.5% SWE increased the yield, nutrient contents, and total soluble solids (TSS) of the four onion cultivars whereas 3% SWE, the highest concentration, increased ascorbic acid in different onion cultivars.

Keywords: ascorbic acid; biostimulants; *Allium cepa*; Phulkara; Nasarpuri; Lambada and Red Bone

---

1. Introduction

Onion (*Allium cepa* L.) belongs to the family Amaryllidaceae. It originated in the Mediterranean region of southwest Asia. More than 800 species of onion are cultivated in the tropical, temperate, and sub-temperate zones of the world [1,2]. Onion is cultivated worldwide due to its economic importance, taste and health-supporting attributes [3]. In the recent past, China, India, the USA, and Turkey were the leading onion producing countries [4]. In Pakistan, onion is grown over an area of 147 thousand acres with an annual production of 1981 thousand tons [5].

Onion has an unbranched root system; it requires a high amount of fertilizers for optimum yield [6]. Chemical fertilizers are commonly used by farmers because of the high concentration of nutrients. Frequent use of chemical fertilizers is a risks to human health and is also a threat to the environment [7]. Increased water eutrophication, nitrate accumulation, heavy metal accumulation and toxicity, and release of nitrogen and sulfur oxides in the air are major environmental concerns. In recent years, rapid emission of greenhouse gases from fertilizers has become a major concern [8]. Application of organic amendments is an environmentally friendly approach to improve crop production and soil sustainability [9]. However there is also some risk of xenobiotic contaminated in manure used to fertilize agricultural soils [10–12].

Biostimulants are emerging as safe alternatives to some chemical fertilizers due to their modulating hormonal, i.e., 'auxin-like, gibberellin-like', effects. They are available in a variety of formulations with varying ingredients. Biostimulants are usually classified into three major groups that includes humic substances (HS), hormone-containing products (HCP), and amino acid-containing products (AACP). Seaweed extracts belong to HCP, contain prominent amounts of active plant growth substances such as auxins, cytokinins, and their derivatives [13]. These growth hormones interact with the biochemical and physiological mechanisms of plants, thus improving crop productivity [14,15].

Moreover, the use of biostimulants may also decrease the fertilizer requirement by enhancing the assimilation of micro- and macro-nutrients [16,17]. Biostimulant benefits have been mainly attributed to the presence of phytohormones and organic molecules [18] such as betaines [18,19]. Such organic compounds are rich in carbohydrates and amino acids and vitamins that increase phenolic compounds [19–21]. Significant improvement in phenolic content approximates the antioxidant activity which decreases reactive oxygen species thus, increase growth and productivity of crops [19,22].

Among biostimulants, seaweed extract (SWE) collected from *Ulva lactuca, Caulerpa sertularioides, Padina gymnospora,* and *Sargassum liebmannii* is the most commonly used amendment [15]. It is mainly comprised of algal extract, glycine betaines, amino acids, and protein hydrolysates. These compounds are very efficient in terms of improving plant growth and productivity [13]. SWE derived from sources like *Ascophyllum nodosum, Ecklonia maxima, Macrocystis pyrifer,a* and *Durvillea potatorum* are also used [20]. SWE extract based on *Ascophyllum nodosum* is enriched in phytohormones, i.e., auxins, gibberellins, cytokinins, abscisic acid, and brassinosteroids [23]. These growth regulators improve growth of plants and enrich overall quality attributes of crops [24]. Generally, all of these could cause a synergistic effect on plant growth via affecting internal metabolic activity although activity and mode of action of these molecules [25,26].

Seaweed extract derived from *Ascophyllum nodosum* along with its co-products is being increasingly utilized to enhance growth and productivity of plants [27] via enhancing seed germination, contributing to vegetative growth and modulating reproductive behavior in tomato [28].

Ascophyllum nodosum extract is comprised of different active substances, i.e., 15–25%, 15–30% alginic acid, 0–10% laminaran, 5–10% mannitol, 4–10% fucoidan, 5–10% protein, 2–7% fats, 2–10% tannins, 0.5–0.9% magnesium, 0.01–1.0% iodine, 2–3% potassium, 3–4% sodium, 0.002–0.006% glycine betaine, and 70–85% water [29]. Betaines and mannitol in seaweed extract helps plants to survive under stress conditions by improving osmotic adaptability [18,30]. However, the variable and multifaceted constituent of these mixtures have made it very effective in promoting growth, yield, and quality attributes of barley (*Hordeum vulgare*) seedlings [31], soybean (*Glycine max*) [32], and cotton (*Gossypium hirsutum*) [33] etc.

The current study was designed to examine the effect of SWE on productivity and quality of onion cultivars. There are few reports on the influence of SWE on onion cultivars, the present study aimed to explore the effective application rate of SWE for onion. We hypothesized that seaweed extract could be an effective biostimulant to improve productivity and quality of onion.

2. Materials and Methods

2.1. Plant Material and Culture Conditions

A field research trial was conducted for one year at the vegetable research farm, Institute of Horticultural Sciences (31.4303° N, 73.0672° E), University of Agriculture, Faisalabad, Pakistan. The physico-chemical characteristics of the sandy loam soil of the planting site are given in Table 1; varies with the land use [34–36]. The field was thoroughly ploughed and harrowed to make ridges and furrows at a 60 cm spacing on finely pulverized soil. Healthy and disease-free seeds of four onion cultivars, 'Phulkara', 'Nasarpuri', 'Lambada', and 'Red Bone' were sown on raised beds in rows 10.2–12.7 cm apart and were lightly covered with soil. To attain uniform seedling growth, necessary cultural and agronomic practices (weeding, irrigation, nutrition and pest management) were used for 45 days until seedling maturity (emergence of 3 to 4 true leaves). The seedling bed was irrigated one day before uprooting and transplanting of seedlings. This helps in proper uprooting with minimal damage and better establishment of seedlings in the field. Seedlings were transplanted at the a spacing of 10 cm in row by 25 cm between rows. Seedling establishment was observed regularly with re-transplanting in case of any dead, weak/injured seedlings.

Table 1. Pre-experiment soil and water characteristics.

Soil	Units	Value	Water	Units	Value
Texture	Loam		pH	-	7.25
pH	-	8.12	Conductivity	$\mu S \cdot cm^{-1}$	941
EC*e*	$dS \cdot m^{-1}$	2.19	Carbonates	$meq \cdot L^{-1}$	0.00
Organic matter	%	0.70	Bicarbonates	$meq \cdot L^{-1}$	0.75
Organic N	%	0.035	Chlorides	$meq \cdot L^{-1}$	1.55
Available P	$mg \cdot kg^{-1}$	6.79	Ca + Mg	$meq \cdot L^{-1}$	9.10
Extractable K	$mg \cdot kg^{-1}$	135	SAR	-	1.51

Abbreviations: EC*e* = Electrical Conductivity of soil extract; TSS = Total Soluble Solids; SAR = Sodium Absorption Ratio; Max = Maximum; Min = Minimum; meq = milliequivalent.

2.2. Collection and Application of Seaweed Extract

The SWE Wokozim (Jaffer Agro Services Private Ltd.) was acquired and, based on preliminary work identifying the optimum concentrations, diluted in water and applied at 0% (control or water only), 0.5%, 1%, 2% and 3% on onion plants. The *Ascophyllum nodosum* extract was characterized as a mixture of cytokinins, auxins, and betaines. As per the company description, the product can be applied as a foliar spray on all parts of plants or can be applied in irrigation water. Three foliar sprays were applied at two-week intervals starting after seedling growth indicated they were established. Each plant receiving approximately 20 mL of solution. The spray was applied to the abaxial and adaxial surface of leaves with complete coverage of plant with a commercial hand-held sprayer.

2.3. Fertilizer and Irrigation

As per production recommendations of the Punjab Agriculture Department, N, P, and K were applied at the rate of 125: 87.5: 87.5 kg ha^{-1}. The first three irrigations of 2.5 cm were applied at 8 days interval. Afterwards, the onion crop was irrigated as required [37].

2.4. Harvesting and Data Collection

Plants were harvested at the time of maturity (>75% of bulb necks fall over). Vegetative growth attributes were immediately measured including, plant height, leaf (consist of a sheath at the bottom which attaches around the stem at a node) and blade length (top of the sheath to blade tip), width and fresh biomass, root length, and total fresh biomass. Among reproductive parameters, bulb weight, bulb and neck diameter (at the narrowest point), total plant dry biomass (tissues were held for 24 h in an oven at 100 °C), and fresh bulb yield hectare^{-1} was calculated. Neck was separated at 2 cm height from the bulb. For moisture determination drying was done in the oven at 100 °C. The difference between fresh and dry weight was used to calculate percent moisture.

2.5. Total Soluble Solids (°Brix)

The total soluble solids (TSS) (°Brix) of bulb tissue was determined with a digital refractometer (ATAGO, RX-5000 Japan).

2.6. Mineral Nutrient and Ascorbic Acid Analyses

Nitrogen, phosphorus, and potassium contents of bulbs were determined according to Chapman and Parker [38]. For N analyses, dry onions were digested with sulfuric acid followed by Kjeldahl distillation. For P and K determination, samples were digested with a di-acid ($HClO_4$:HNO_3 at a 2:1 ratio) mixture. Finally, a colorimetric method was used and concentrations were assessed at 430 nm using a spectrophotometer for P analysis. For the development of yellow color for P quantification, ammonium molybdate and ammonium metavanadate chemicals were used. For K determination, digested samples were run on a flame photometer after filtration [38]. For ascorbic acid quantification, the dichlorophenolindophenol (DCPIP) method was used. Initially, 10 g onion sample was taken and its juice was taken with 2.5 mL of 20% metaphosphoric acid and diluted with distilled water to 100 mL which developed blue color as per presence of ascorbic acid. Finally, 10 mL of prepared extract was titrated with 0.1% DCPIP until the blue color disappeared and the solution became colorless [39].

2.7. Statistical Analysis

The experiment was laid out in a Randomized Complete Block Design (RCBD) with nine replications of each cultivar by SWE concentration. Statistix 8.1 computer software was used to analyze the data. Data pass the test for homogeneity and then analysis of variance (ANOVA) was used to determine the overall significance of data at $p \leq 0.05$. Treatments were compared using Fisher's least significant difference (LSD) test at $p \leq 0.05$ [40].

3. Results

3.1. Leaf-Blade Length and Plant Height

SWE application had significant ($p \leq 0.05$) effects on leaf blade length of the different onion cultivars. A maximum increase of 65% was observed for 'Phulkara', 35.7% for 'Red Bone', 25.0% for 'Lambada' and 17% for 'Nasar puri' at 0.5% SWE as compared to the control (Figure 1A). Application of 0.5 and 1% SWE significantly increased leaf blade length in all cultivars. However, application of 3% SWE increased leaf blade length in 'Phulkara' and 'Red Bone' but decreased leaf blade length in 'Nasar puri' and 'Lambada' as compared to the control.

Treatment with 0.5% SWE resulted in the most increase in plant height of all four cultivars. The increase in plant height was 29.7% in 'Phulkara', 25.4% in 'Lambada', 19.9% in 'Red Bone', and 17.9% in 'Nasar puri' as compared to control (Figure 1B). Generally, no significant change in plant height was noted when SWE concentration was increased from 0.5% and 1%. SWE at 2% also significantly enhanced plant height in 'Lambada' and 'Phulkara' but did not affect 'Nasar puri' and 'Red Bone' as compared to control (Figure 1B).

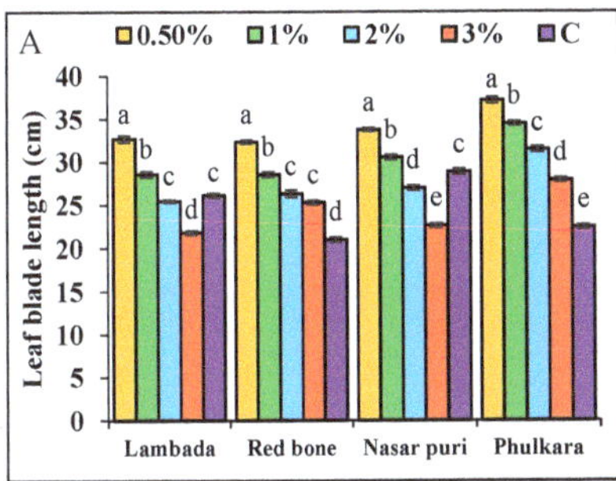
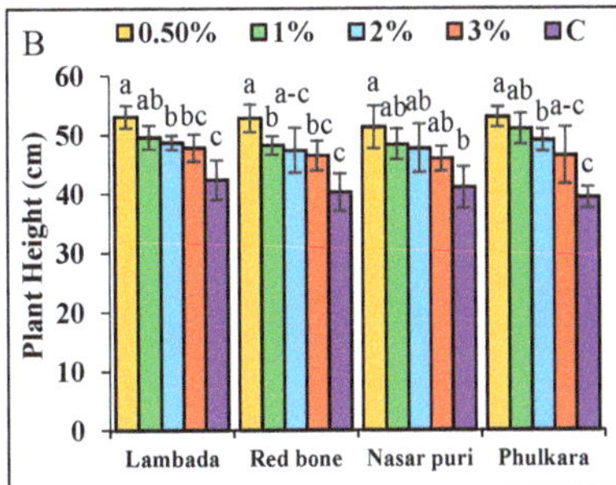

Figure 1. Leaf-blade length (**A**) and plant height (**B**) of onion cultivars as affected by foliar application of 0 (C = control), 0.5, 1, 2, or 3% seaweed extract. Values are means of nine replicates ± standard error. Means with different letters are significantly different from each other compared by least significant difference (LSD) at $p \leq 0.05$.

3.2. Leaf Length and Width

Application of SWE significantly ($p \leq 0.05$) improved leaf length and width of onion cultivars. A maximum increase of 38.6% in 'Nasar puri', 25.1% in 'Lambada', 14.3% in 'Red Bone', 13.3% in 'Phulkara' was noted in leaf length at 2% SWE over the control (Figure 2A). In 'Lambada', 1% and 3% SWE showed similar effects with each other and were both greater than the control. In 'Red Bone', 1% and 3% were similar to the control. However, 0.5% SWE caused a significant reduction in leaf length of 'Red Bone'. For 'Nasar puri', both 1 and 3% SWE also significantly improved leaf length, but 0.5% SWE did not differ from the control. Application of 3% SWE significantly enhanced but 0.5 and 1% cause a reduction in leaf length of 'Phulkara'.

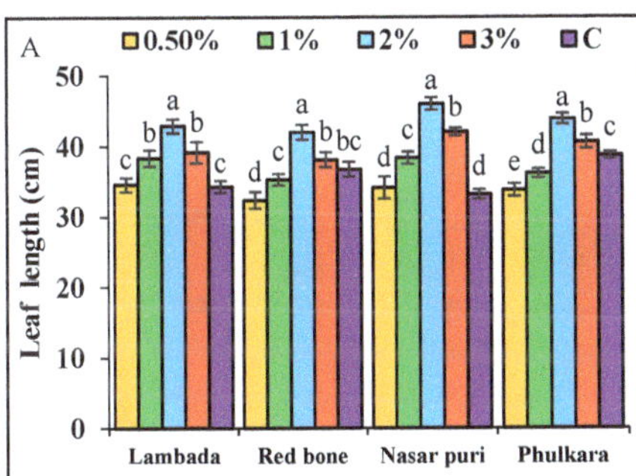
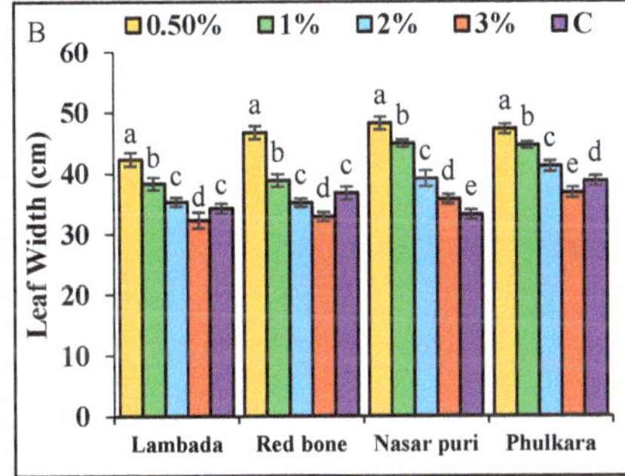

Figure 2. Leaf length (**A**) and width (**B**) in onion cultivars as affected by foliar application of 0 (C = control), 0.5, 1, 2, or 3% seaweed extract. Values are means of nine replicates ± standard error. Means with different letters are significantly different from each other compared by LSD at $p \leq 0.05$.

Leaf width showed a maximum increase of 38.6% in 'Nasar puri', 35.2% in 'Nasar puri', 15.2% in 'Phulkara' and 11.9% in 'Lambada' at 0.5% SWE as compared to the control. Application of 0.5% and 1% SWE both increased leaf width in the four cultivars (Figure 2B). SWE at 2% did not differ significantly from the control in 'Red Bone' and 'Lambada' but remained significantly greater in 'Phulkara' and 'Nasar puri'. SWE at 3% caused a significant reduction in leaf width of 'Phulkara', 'Red Bone' and 'Lambada' but increased leaf width in 'Nasar puri' over control.

3.3. Root Length and Weight

Application of SWE significantly ($p \leq 0.05$) affected onion root length and weight. A maximum increase of 81.0% in 'Phulkara', 70.0% in 'Nasar puri', 40.3% in 'Red Bone' and 9.2% in 'Lambada' root length at 0.5% SWE were observed (Figure 3A). In 'Lambada', 0.5% and 1%, SWE significantly increased root length, but 2% and 3% SWE decreased it. Application of 0.5%, 1% and 2% SWE remained statistically alike but 3% decreased root length over control in 'Red Bone'. In 'Nasar puri', all SWE rates enhanced root length as compared to the control. However, 0.5%, 1% and 2% SWE significantly improved but 3% SWE did not affect root length in 'Phulkara'.

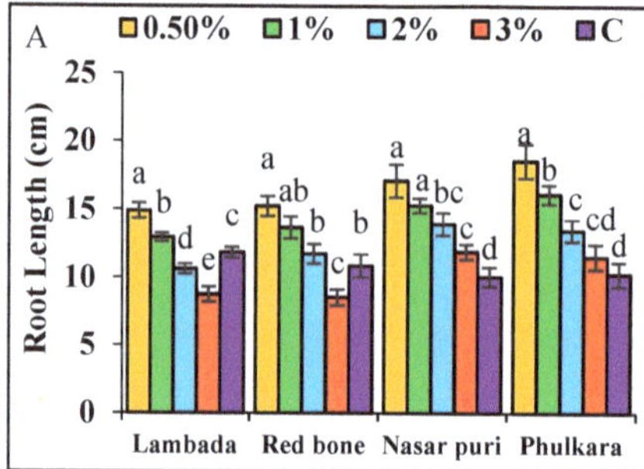
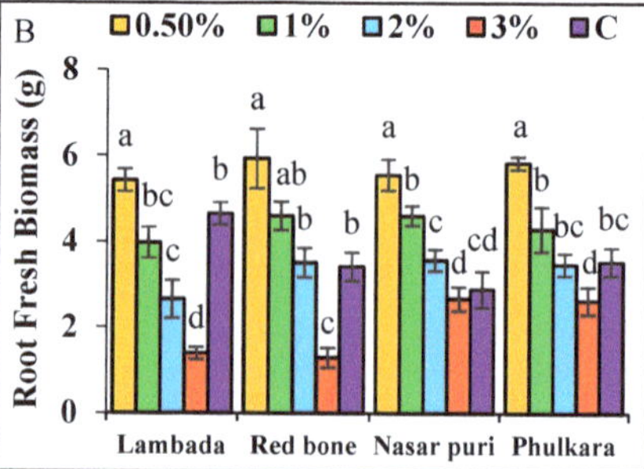

Figure 3. Root length (**A**) and fresh biomass (**B**) in onion cultivars as affected by foliar application of 0 (C = control), 0.5, 1, 2, or 3% seaweed extract. Values are means of nine replicates ± standard error. Means with different letters are significantly different from each other compared by LSD at $p \leq 0.05$.

A maximum increase of 92.7% in 'Nasar puri', 73.4% in 'Red Bone', 65.2% in 'Phulkara' and 16.6% in 'Lambada' root weight at 0.5% SWE was observed as compared to the control (Figure 3B). In 'Lambada', no significant change in root weight was observed at 1% SWE. However, 2% and 3% SWE decreased root weight in 'Lambada'. For root weight in 'Red Bone', 1% and 2% SWE were statistically alike but 3% caused a significant reduction as compared to the control. Application of 1% SWE in 'Nasar puri' significantly enhanced root weight from control but 2% and 3% SWE did not differ from the control. No significant change was observed with 1% or 2% SWE. However, 0.5% SWE significantly increased root weight, while 3% SWE decreased root weight from control in 'Phulkara'.

3.4. Bulb and Neck Diameter

Application of SWE significantly ($p \leq 0.05$) affected bulb diameter and neck diameter of the onion cultivars (Figure 4A). Bulb diameter showed a maximum increase of 23.9%, 9.0% and 18.3% in bulb diameter in 'Lambada', 'Nasar puri', and 'Red Bone' respectively as compared to the control at 0.5% SWE. However, in 'Phulkara' a maximum increase of bulb diameter of 13.2% was noted where 3% SWE was applied. In 'Lambada' all levels of SWE 0.5, 1%, 2%, and 3% caused significant improvement in bulb diameter as compared to the control. Application of 0.5 and 1% SWE were statistically alike but only 0.5% SWE caused a significant increase in 'Red Bone' bulb diameter. However, 3% of SWE caused a significant reduction in bulb diameter of 'Red Bone'. In 'Nasar puri', 0.5 and 3% SWE showed significant positive increases but 2% SWE did not differ from the control. Furthermore, the application of 1% SWE negatively affected bulb diameter in 'Nasar puri'. In 'Phulkara', except for 3% SWE, no application rate improved bulb diameter. However, 0.5 and 1% SWE significantly decreased bulb diameter weight in 'Phulkara'.

Maximum increases of 71.8%, 62.9%, 48.0%, 38.9% were noted in neck diameter of 'Red Bone', 'Red Bone', 'Lambada' 'Nasar puri', and 'Phulkara' at 0.5% SWE as compared to the control (Figure 4B). All application rates of SWE significant improved neck diameter of 'Lambada' and 'Red Bone'. In 'Nasar puri' and 'Phulkara', 0.5% and 1% SWE caused a significant increase, 1% SWE did not differ from the control, and 3% SWE decreased neck diameter.

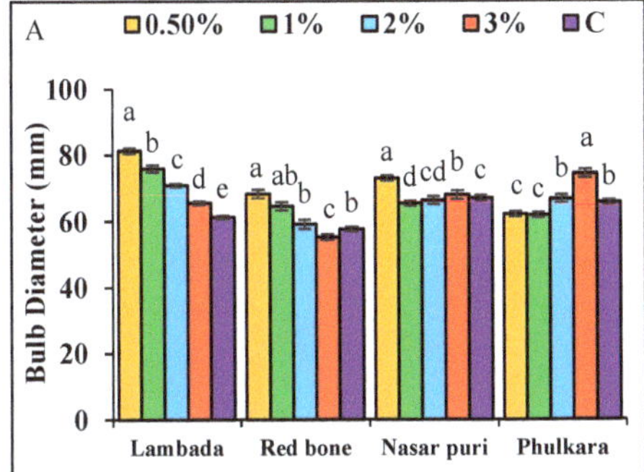
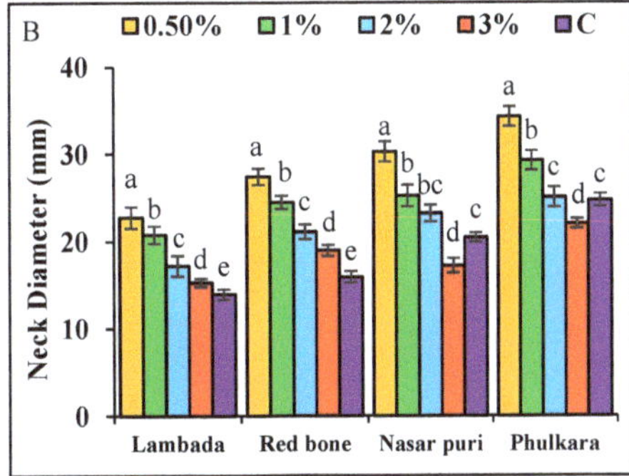

Figure 4. Bulb diameter (**A**) and neck diameter (**B**) in onion cultivars as affected by foliar application of 0 (C = control), 0.5, 1, 2, or 3% seaweed extract. Values are means of nine replicates ± standard error. Means with different letters are significantly different from each other compared by LSD at $p \leq 0.05$.

3.5. Bulb Dry Weight and Moisture Content

Application of SWE significantly ($p \leq 0.05$) affected bulb dry matter and neck moisture of different onion cultivars (Figure 5A). A maximum increase in bulb dry matter of 59.3% in 'Red Bone', 49.4% in 'Phulkara', 21.8% in 'Lambada' and 17.2% in 'Nasar puri' was noted at 0.5% SWE. Application of 0.5% and 1% SWE resulted in significant improvement in bulb dry matter of 'Lambada', 'Red Bone', 'Nasar puri' and 'Phulkara'. SWE at 2% did not differ from the control but 3% SWE caused a significant decrease in bulb dry matter of 'Lambada', 'Red Bone', 'Nasar puri' and 'Phulkara'.

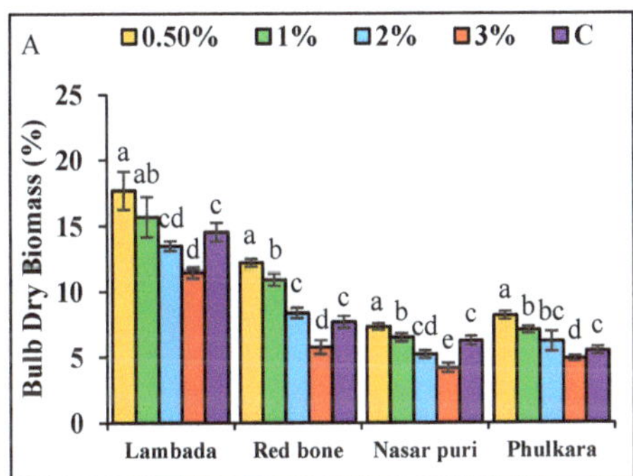
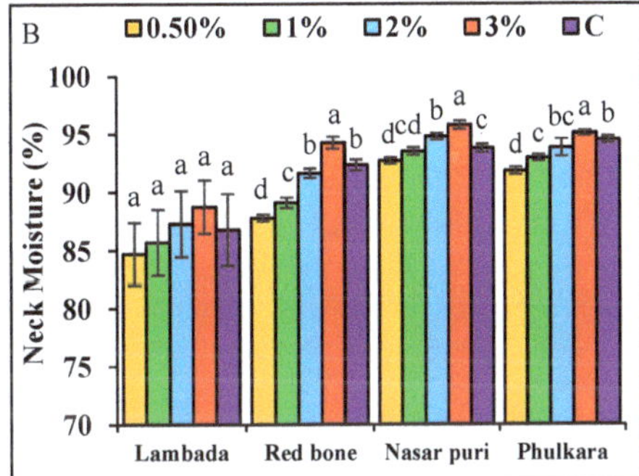

Figure 5. Bulb dry biomass (**A**) and neck moisture (**B**) in onion cultivars as affected by foliar application of 0 (C = control), 0.5, 1, 2, or 3% seaweed extract. Values are means of nine replicates ± standard error. Means with different letters are significantly different from each other compared by LSD at $p \leq 0.05$.

A maximum increase of 2.19%, 2.09% and 0.63% in neck moisture was noted in 'Nasar puri', 'Red Bone' and 'Phulkara' at 3% SWE as compared to the control (Figure 5B). No application rate of SWE differed from the control for 'Lambada' neck moisture. It was noted that 1% SWE did not differ from the control in 'Red Bone' and 'Phulkara' but was significantly different in 'Nasar puri'. Application of 2–3% SWE significantly decreased neck moisture in 'Red Bone' and 'Phulkara' but only 3% caused a reduction in 'Nasar puri'.

3.6. Bulb Weight and Yield

Application of SWE significantly ($p \leq 0.05$) affected bulb weight and yield of the onion cultivars (Figure 6A). A maximum increase in bulb weight of 5.8, 5.4, 2.4, and 2.0% in 'Lambada', 'Red Bone', 'Phulkara', and 'Nasar puri' at 0.5% SWE was observed, respectively. SWE at 0.5 and 1% significantly enhanced bulb weight, 2% SWE had no effect but 3% SWE caused a reduction of 'Lambada'. In 'Red Bone', 0.5% SWE caused a significant improvement in bulb weight but 1%, 2%, 3% SWE did not differ significantly from the control. In 'Nasar puri' and 'Phulkara', 0.5% SWE significantly improved bulb weight, 1 and 2% had no effect but 3% SWE caused a significant reduction in bulb weight.

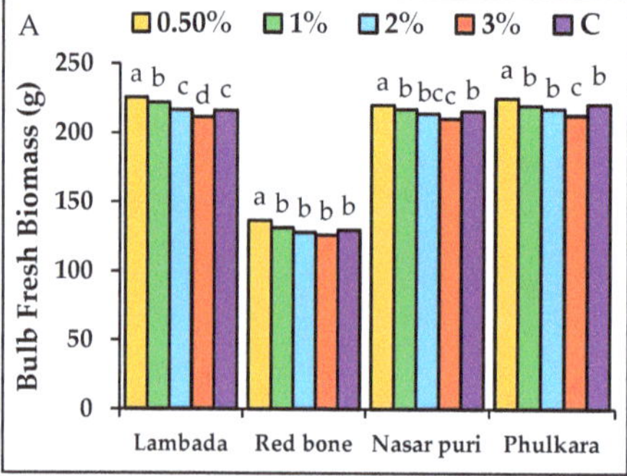
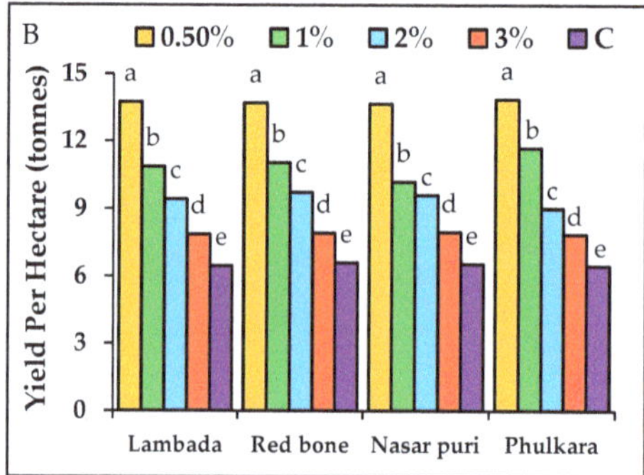

Figure 6. Bulb fresh biomass (**A**) and yield per hectare (**B**) in onion cultivars as affected by foliar application of 0 (C = control), 0.5, 1, 2, or 3% seaweed extract. Values are means of nine replicates ± standard error. Means with different letters are significantly different from each other compared by LSD at $p \le 0.05$.

Regarding yield per hectare, application of 0.5%, 1%, 2%, and 3% SWE caused significant increases 'Lambada', 'Red Bone', 'Phulkara', and 'Nasar puri' (Figure 6B).

3.7. Total Soluble Solids

Total soluble solids and ascorbic acid showed significant ($p \le 0.05$) positive responses to SWE application (Figure 7A). SWE at 0.5% gave a maximum increase in TSS of 175% in 'Lambada', 137% in 'Red Bone', 111% in 'Phulkara' and 30% in 'Nasar puri' over the control. On average, all application rates of SWE showed significant improvements in TSS of 'Lambada', 'Red Bone' and 'Phulkara'. However, in 'Nasar puri', 2 and 3% SWE caused significant reductions in TSS. For all onion cultivars, increasing application rates of SWE also enhanced ascorbic acid content (Figure 7B). Maximum increases of 58.8%, 52.9%, 58.8% and 58.3% in 'Lambada', 'Red Bone', 'Nasar puri', and 'Phulkara', respectively, were observed.

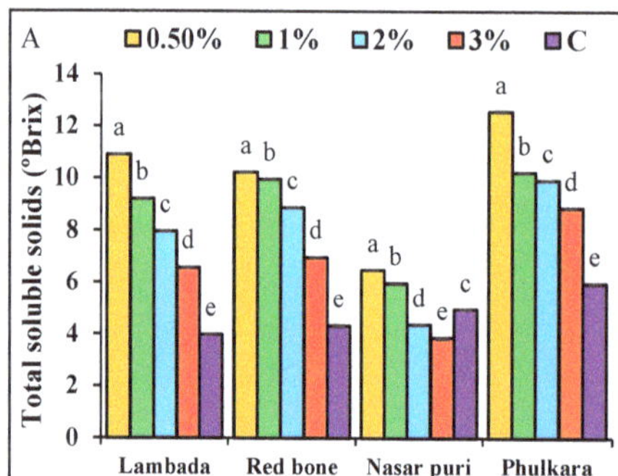
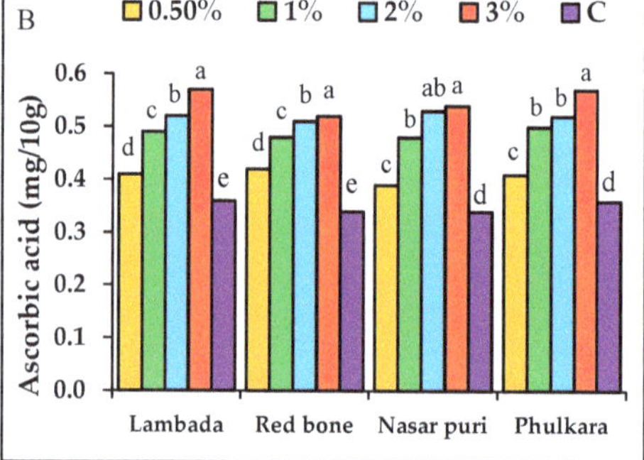

Figure 7. Total soluble solids (**A**) and ascorbic acid (**B**) in onion cultivars as affected by foliar application of 0 (C = control), 0.5, 1, 2, or 3% seaweed extract. Values are means of nine replicates ± standard error. Means with different letters are significantly different from each other compared by LSD at $p \le 0.05$.

3.8. Nitrogen, Phosphorus and Potassium Contents

Application of SWE significantly ($p \leq 0.05$) affected N, P and K contents of the onion cultivars (Figure 8A). SWE at 0.5% resulted in a maximum increase of 83%, 127%, and 150% in N content of 'Lambada', 'Red Bone', and 'Phulkara', respectively, over the control. However, in 'Nasar puri', a greater increase in N content was observed at 3% SWE. All other application rates of SWE also significantly increased N content of the onion cultivars. For P content, again 0.5% SWE resulted in the greatest improvement of 57.6% in 'Lambada', 57.5% in 'Phulkara' and 47.3% in 'Red Bone' as compared to the control (Figure 8B). However, 1% SWE also caused a 38.3% increase in P content of 'Nasar puri'. Increasing levels of SWE gave decreasing trends in P content of all onion cultivars. For K content, 0.5% SWE showed a maximum increase of 26.3% in 'Red Bone', 57.6% in 'Lambada', 29.4% in 'Nasar puri' and 42.8% in 'Phulkara' compared to the control (Figure 8C). Application of 1% SWE also improved P content, but 2% SWE had no effect on K content of any onion cultivar (Figure 8C). However, 3% SWE caused a significant reduction in K content of 'Lambada', 'Red Bone', and 'Phulkara'.

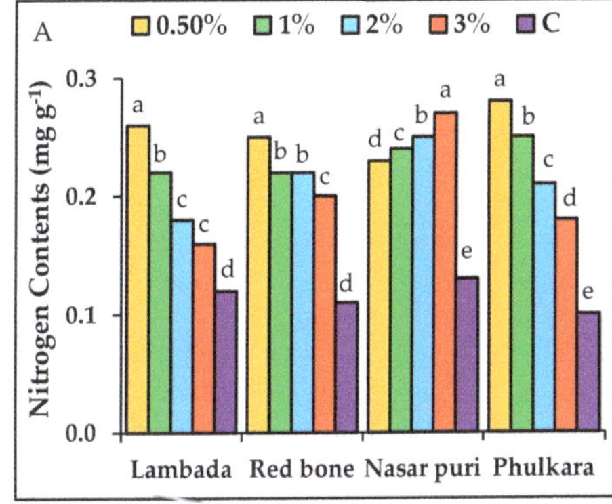
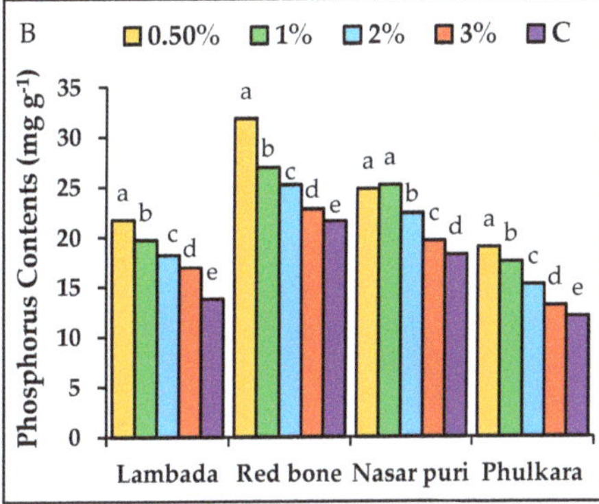
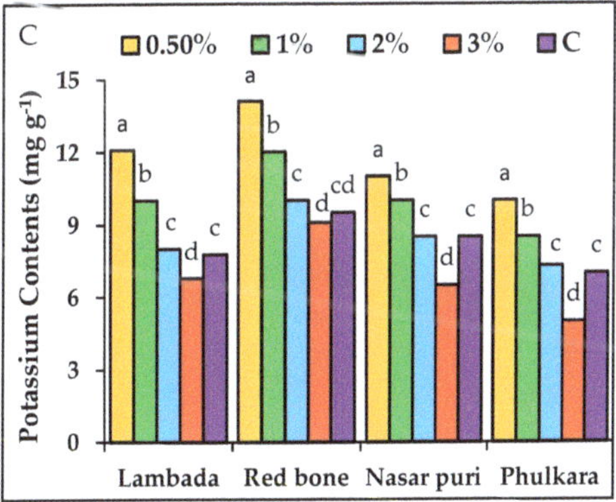

Figure 8. Nitrogen (**A**), phosphorus (**B**) and potassium (**C**) contents per g dry biomass in onion cultivars as affected by foliar application of 0 (C = control), 0.5, 1, 2, or 3% seaweed extract. Values are means of nine replicates ± standard error. Means with different letters are significantly different from each other compared by LSD at $p \leq 0.05$.

4. Discussion

In the present study, the results showed that foliar application of seaweed extract significantly enhanced different vegetative characters, i.e., leaf length, blade width, and plant height. This increment in leaf length and blade width could be ascribed to the role of seaweed enriched in auxin and cytokinins. These hormones affect metabolic activities predominantly through cell division and elongation [41].

Moreover, seaweed extract comprised of carbohydrates, mainly oligosaccharides, may control plant growth by affecting N assimilation and basal metabolism [42] by altering gene expression [43] whereas the number of leaves per plant may be influenced by prevailing climatic conditions, internal factors, and the interaction of both [44]. Vigor and health of leaves are chiefly chlorophyll dependent, as under stress conditions degradation of chlorophyll can lead to senescence of leaves [45]. SWE may prevent this chlorophyll degradation [46] due to the presence of betaines that protect the thylakoid membrane by regulating osmotic adjustment and enhancing ion homeostasis [47].

The present results also indicated that SWE application led to an increase in root length which can be ascribed to alginate oligosaccharide-induced expression of an auxin-related gene leading to higher auxin concentrations, thus promoting root formation and elongation [48]. Moreover, it could also be associated with SWE-modified absorption and localization of auxin and cytokinins that initiates lateral and adventitious root development along with heavier root biomass [49]. However, increased root growth could also be due to SWE- induced uptake and utilization of mineral nutrients [50], particularly N and S content [51]. SWE enriched in vitamins A, B and E [52] may help to enhance reproductive characters of onion as vitamin E is known to have a significant effect on bulb and neck diameter, and fresh and dry weight [53].

The present results also showed that all concentrations of seaweed exhibited higher bulb weight. This could be ascribed to auxin supplied by the seaweed which in turn enhanced cell division, elongation, and differentiation in addition to enhanced uptake of higher proteins and nucleic acid reserves eventually ensuring higher bulb weight [54]. Nutritional contents of onion were improved in SWE-applied treatments as the presence of glycine betaine, a constituent of SWE, could have led to enhanced phenolic compound synthesis [55] that has also been correlated to a positive relation between TSS, sweetness and ascorbic acid contents of fruits [56]. Total soluble solids depend upon transported ions and organic solutes which are converted into glucose inside fruit [57] where SWE application enhances glucose biosynthesis contributing to improved TSS content.

SWE extract application led to enhanced uptake and retention of nutrients in all onion cultivars. Jannin et al. [51] also concluded that SWE extracts positively led to the increment of root growth that ultimately enhanced the uptake and accumulation of elemental N content in bulbs. Similar results were reported by Halpern et al. [50] where seaweed extract significantly increased the uptake of N, P, and K. Also, SWE extract containing oligosaccharides resulted in increased photosynthesis, basal metabolism, cell division and altered metabolic pathways for increased uptake and better assimilation of nitrogen [58].

5. Conclusions

Foliar application of SWE positively influenced growth, yield, and quality attributes of in four onion cultivars. In general, the lowest rate used, 0.5% of SWE, had a significant impact on the onion cultivars nutrient content, yield, and TSS. However, the application of the highest rate of 3% SWE was not efficacious for promoting growth and yield attributes other than increased ascorbic acid. More investigations are suggested under different soil and climatic conditions to determine the best application rate of seaweed extract in field production of onions.

Author Contributions: M.A. designed and supervised experiment and wrote manuscript. J.A. conducted research, collected data and wrote manuscript. M.Z.-u.-H. and A.A.R. assisted in methodology validation and manuscript writing. R.I.K. and M.S. assisted in preparation of manuscript. S.D. and R.D. conducted statistical analyses, reviewed manuscript. All authors have read and agreed to the published version of the manuscript.

Funding: This research received no external funding.

Acknowledgments: We are highly thankful to Anwar (Pak Agri Gujranwala, Punjab, Pakistan) for providing Red bone and Iqbal (More green seeds company Lahore, Punjab, Pakistan) for providing Lambada Cultivars of onion.

Conflicts of Interest: The authors declare no conflict of interest.

References

1. Fritsch, R.M.; Friesen, N. Evolution, domestication and taxonomy. In *Allium Crop Science: Recent Advances*; Rabinowitch, H.D., Currah, L., Eds.; CABI: New York, NY, USA, 2002; pp. 5–30.
2. Wani, A.H. Management of black mold rot of onion. *Mycopath* **2011**, *9*, 43–49.
3. Suleria, H.A.R.; Butt, M.S.; Anjum, F.M.; Saeed, F.; Khalid, N. Onion: Nature Protection Against Physiological Threats. *Crit. Rev. Food Sci. Nutr.* **2015**, *55*, 50–66. [CrossRef] [PubMed]
4. Kao, M.C.J.; Gesmann, M.; Gheri, F. FAOSTAT: Download Data from the FAOSTAT Database of the Food and Agricultural Organization (FAO) of the United Nations. Available online: https://cran.r-project.org/web/packages/FAOSTAT/index.html (accessed on 15 September 2019).
5. Pakistan Bureau of Statistics. *Agriculture: Onion Production*; Pakistan Economic Survey 2017-18; Government of Pakistan, Ministry of Finance: Islamabad, Pakistan, 2018. Available online: http://www.finance.gov.pk/survey/chapters_18/02-Agriculture.pdf (accessed on 5 August 2019).
6. Rizk, F.A.; Shaheen, A.M.; Abd El-Samad, E.A.; Sawan, O.M. Effect of different nitrogen plus phosphorus and sulphur fertilizer levels on growth, yield and quality of onion (*Ailium cepa* L.). *J. Appl. Sci. Res.* **2012**, *8*, 3353–3361.
7. Ali, M.; Khan, N.; Khan, A.; Rafeh Ullah, A.N.; Khan, M.W.; Khan, K.; Farooq, S.; Rauf, K. Organic manures effect on the bulb production of onion cultivars under semiarid condition. *Pure Appl. Biol.* **2018**, *7*, 1161–1170. [CrossRef]
8. Savci, S. Investigation of Effect of Chemical Fertilizers on Environment. *APCBEE Procedia* **2012**, *1*, 287–292. [CrossRef]
9. Brtnicky, M.; Dokulilova, T.; Holatko, J.; Pecina, V.; Kintl, A.; Latal, O.; Vyhnanek, T.; Prichystalova, J.; Datta, R. Long-term effects of biochar-based organic amendments on soil microbial parameters. *Agronomy* **2019**, *9*, 747. [CrossRef]
10. Molaei, A.; Lakzian, A.; Haghnia, G.; Astaraei, A.; Rasouli-Sadaghiani, M.; Ceccherini, M.T.; Datta, R. Assessment of some cultural experimental methods to study the effects of antibiotics on microbial activities in a soil: An incubation study. *PLoS ONE* **2017**, *12*, e0180663.
11. Molaei, A.; Lakzian, A.; Datta, R.; Haghnia, G.; Astaraei, A.; Rasouli-Sadaghiani, M.; Ceccherini, M.T. Impact of chlortetracycline and sulfapyridine antibiotics on soil enzyme activities. *Int. Agrophys.* **2017**, *31*, 499–505. [CrossRef]
12. Meena, R.S.; Kumar, S.; Datta, R.; Lal, R.; Vijayakumar, V.; Brtnicky, M.; Sharma, M.P.; Yadav, G.S.; Jhariya, M.K.; Jangir, C.K. Impact of Agrochemicals on Soil Microbiota and Management: A Review. *Land* **2020**, *9*, 34. [CrossRef]
13. Du Jardin, P. Plant biostimulants: Definition, concept, main categories and regulation. *Sci. Hortic.* **2015**, *196*, 3–14. [CrossRef]
14. Calvo, P.; Nelson, L.; Kloepper, J.W. Agricultural uses of plant biostimulants. *Plant Soil* **2014**, *383*, 3–41. [CrossRef]
15. Hernández-Herrera, R.M.; Santacruz-Ruvalcaba, F.; Ruiz-López, M.A.; Norrie, J.; Hernández-Carmona, G. Effect of liquid seaweed extracts on growth of tomato seedlings (Solanum lycopersicum L.). *J. Appl. Phycol.* **2014**, *26*, 619–628. [CrossRef]
16. Nardi, S.; Carletti, P.; Pizzeghello, D.; Muscolo, A. Biological activities of humic substances, in biophysicochemical processes involving natural nonliving organic matter in environmental systems. In *Fundamentals and Impact of Mineral-Organic-Biota Interactions on the Formation, Transformation, Turnover, and Storage of Natural Nonliving Organic Matter (NOM)*; Senesi, N., Xing, B., Huang, P.M., Eds.; John Wiley: Hoboken, NJ, USA, 2009.
17. Ertani, A.; Nardi, S.; Altissimo, A. Long-term research activity on the biostimulant properties of natural origin compounds. *Acta Hortic.* **2012**, *1009*, 181–187.
18. Battacharyya, D.; Babgohari, M.Z.; Rathor, P.; Prithiviraj, B. Seaweed extracts as biostimulants in horticulture. *Sci. Hortic.* **2015**, *196*, 39–48. [CrossRef]
19. Blunden, G.; Morse, P.F.; Mathe, I.; Hohmann, J.; Critchley, A.T.; Morrell, S. Betaine yields from marine algal species utilized in the preparation of seaweed extracts used in agriculture. *Nat. Prod. Commun.* **2010**, *5*, 581–585. [CrossRef]

20. Khan, W.; Rayirath, U.P.; Subramanian, S.; Jithesh, M.N.; Rayorath, P.; Hodges, D.M.; Critchley, A.T.; Craigie, J.S.; Norrie, J.; Prithiviraj, B. Seaweed extracts as biostimulants of plant growth and development. *J. Plant Growth Regul.* **2009**, *28*, 386–399. [CrossRef]

21. Berlyn, G.P.; Russo, R.O. The use of organic biostimulants to promote root growth. *Belowground Ecol.* **1990**, *2*, 12–13.

22. Datta, R.; Baraniya, D.; Wang, Y.-F.; Kelkar, A.; Meena, R.; Yadav, G.; Teresa Ceccherini, M.; Formanek, P. Amino Acid: Its Dual Role as Nutrient and Scavenger of Free Radicals in Soil. *Sustainability* **2017**, *9*, 1402.

23. Stirk, W.A.; Tarkowská, D.; Turečová, V.; Strnad, M.; van Staden, J. Abscisic acid, gibberellins and brassinosteroids in Kelpak®, a commercial seaweed extract made from Ecklonia maxima. *J. Appl. Phycol.* **2014**, *26*, 561–567. [CrossRef]

24. Rouphael, Y.; Colla, G.; Giordano, M.; El-Nakhel, C.; Kyriacou, M.C.; De Pascale, S. Foliar applications of a legume-derived protein hydrolysate elicit dose-dependent increases of growth, leaf mineral composition, yield and fruit quality in two greenhouse tomato cultivars. *Sci. Hortic.* **2017**, *22*, 353–360. [CrossRef]

25. Vernieri, P.; Ferrante, A.; Borghesi, E.; Magnani, G. High quality flowering plants using the biostimulants. *L'Informatore Agrario* **2005**, *16*, 57–60.

26. Wally, O.S.D.; Critchley, A.T.; Hiltz, D.; Craigie, J.S.; Han, X.; Zaharia, L.I.; Abrams, S.R.; Prithiviraj, B. Regulation of Phytohormone Biosynthesis and Accumulation in Arabidopsis Following Treatment with Commercial Extract from the Marine Macroalga Ascophyllum nodosum. *J. Plant Growth Regul.* **2013**, *32*, 324–339. [CrossRef]

27. Cristian Popescu, G.; Popescu, M. Effect of the Brown Alga Ascophyllum Nodosum As Biofertilizer on Vegetative Growth in Grapevine (Vitis Vinifera L.). *Curr. Trends Nat. Sci.* **2014**, *3*, 61–67.

28. Kumari, R.; Kaur, I.; Bhatnagar, A.K. Effect of aqueous extract of Sargassum johnstonii Setchell & Gardner on growth, yield and quality of Lycopersicon esculentum Mill. *J. Appl. Phycol.* **2011**, *23*, 623–633. [CrossRef]

29. Chojnacka, K. Biologically Active Compounds in Seaweed Extracts - the Prospects for the Application. *Open Conf. Proc. J.* **2012**, *3*, 20–28. [CrossRef]

30. Lojkova, L.; Vranová, V.; Formánek, P.; Drápelová, I.; Brtnicky, M.; Datta, R. Enantiomers of Carbohydrates and Their Role in Ecosystem Interactions: A Review. *Symmetry* **2020**, *12*, 470.

31. Yaronskaya, E.; Vershilovskaya, I.; Poers, Y.; Alawady, A.E.; Averina, N.; Grimm, B. Cytokinin effects on tetrapyrrole biosynthesis and photosynthetic activity in barley seedlings. *Planta* **2006**, *224*, 700–709.

32. Szparaga, A.; Kocira, S.; Kocira, A.; Czerwińska, E.; Świeca, M.; Lorencowicz, E.; Kornas, R.; Koszel, M.; Oniszczuk, T. Modification of growth, yield, and the nutraceutical and antioxidative potential of soybean through the use of synthetic biostimulants. *Front. Plant Sci.* **2018**, *871*. [CrossRef]

33. De Silva, R.A.; Santos, J.L.; Oliveira, L.S.; Soares, M.R.S.; Dos Santos, S.M.S. Biostimulants on mineral nutrition and fiber quality of cotton crop. *Rev. Bras. Eng. Agric. Ambient.* **2016**, *20*, 1062–1066. [CrossRef]

34. Danso Marfo, T.; Datta, R.; Vranová, V.; Ekielski, A. Ecotone Dynamics and Stability from Soil Perspective: Forest-Agriculture Land Transition. *Agriculture* **2019**, *9*, 228.

35. Marfo, T.D.; Datta, R.; Pathan, S.I.; Vranová, V. Ecotone Dynamics and Stability from Soil Scientific Point of View. *Diversity* **2019**, *11*, 53. [CrossRef]

36. Yadav, G.; Datta, R.; Imran Pathan, S.; Lal, R.; Meena, R.; Babu, S.; Das, A.; Bhowmik, S.; Datta, M.; Saha, P. Effects of Conservation Tillage and Nutrient Management Practices on Soil Fertility and Productivity of Rice (Oryza sativa L.)–Rice System in North Eastern Region of India. *Sustainability* **2017**, *9*, 1816. [CrossRef]

37. GOP Onion Production Technology. Directorate of Agriculture Information. Available online: http://dai.agripunjab.gov.pk/system/files/OnionPlan2019-20_0.pdf (accessed on 23 September 2019).

38. Chapman, H.D.; Pratt, P.F. *Methods of Analysis For Soils, Plants and Water*; University of California, Division of Agricultural Sciences: Berkeley, CA, USA, 1961.

39. Patrick, A.O.; Fabian, U.A.; Peace, I.C.; Fred, O.O. Determination of Variation of Vitamin "C" Content of Some Fruits and Vegetables Consumed in Ugbokolo After Prolonged Storage. *IOSR J. Environ. Sci.* **2016**, *10*, 17–19. [CrossRef]

40. Steel, R.G.; Torrie, J.H.; Dickey, D.A. *Principles and Procedures of Statistics: A Biometrical Approach*, 3rd ed.; McGraw Hill Book International Co.: Singapore, 1997.

41. Prasad, B.D.; Goel, S.; Krishna, P. In Silico identification of carboxylate clamp type tetratricopeptide repeat proteins in Arabidopsis and rice as putative co-chaperones of Hsp90/Hsp70. *PLoS ONE* **2010**, *5*, 12761. [CrossRef]

42. El-Mohdy, H.L.A. Radiation-induced degradation of sodium alginate and its plant growth promotion effect. *Arab. J. Chem.* **2017**, *10*, 431–438. [CrossRef]
43. Farmer, E.E.; Moloshok, T.D.; Saxton, M.J.; Ryan, C.A. Oligosaccharide signaling in plants. *J. Biol. Chem.* **1991**, *266*, 3140–3145.
44. Ghanem, M.E.; Albacete, A.; Martínez-Andújar, C.; Acosta, M.; Romero-Aranda, R.; Dodd, I.C.; Lutts, S.; Pérez-Alfocea, F. Hormonal changes during salinity-induced leaf senescence in tomato (Solanum lycopersicum L.). *J. Exp. Bot.* **2008**, *59*, 3039–3050. [CrossRef]
45. Noodén, L.D.; Guiamét, J.J.; John, I. Senescence mechanisms. *Physiol. Plant.* **1997**, *101*, 746–753. [CrossRef]
46. Whapham, C.A.; Blunden, G.; Jenkins, T.; Hankins, S.D. Significance of betaines in the increased chlorophyll content of plants treated with seaweed extract. *J. Appl. Phycol.* **1993**, *5*, 231–234. [CrossRef]
47. Tian, F.; Wang, W.; Liang, C.; Wang, X.; Wang, G.; Wang, W. Overaccumulation of glycine betaine makes the function of the thylakoid membrane better in wheat under salt stress. *Crop J.* **2017**, *5*, 73–82. [CrossRef]
48. Zhang, Y.; Yin, H.; Zhao, X.; Wang, W.; Du, Y.; He, A.; Sun, K. The promoting effects of alginate oligosaccharides on root development in Oryza sativa L. mediated by auxin signaling. *Carbohydr. Polym.* **2014**, *113*, 446–454. [CrossRef] [PubMed]
49. Kurepin, L.; Haslam, T.; Lopez-Villalobos, A.; Oinam, G.; Yeung, E. Adventitious root formation in ornamental plants: II. The role of plant growth regulators. *Propag. Ornam. Plants* **2011**, *11*, 161–171.
50. Halpern, M.; Bar-Tal, A.; Ofek, M.; Minz, D.; Muller, T.; Yermiyahu, U. The Use of Biostimulants for Enhancing Nutrient Uptake. *Adv. Agron.* **2015**, *130*, 141–158. [CrossRef]
51. Jannin, L.; Arkoun, M.; Etienne, P.; Laîné, P.; Goux, D.; Garnica, M.; Fuentes, M.; Francisco, S.S.; Baigorri, R.; Cruz, F.; et al. Brassica napus Growth is Promoted by Ascophyllum nodosum (L.) Le Jol. Seaweed Extract: Microarray Analysis and Physiological Characterization of N, C, and S Metabolisms. *J. Plant Growth Regul.* **2013**, *32*, 31–52. [CrossRef]
52. Madhusudan, C.; Manoj, S.; Rahul, K.; Rishi, C.M. Seaweeds: A diet with nutritional, Medicinal and Industrial value. *Res. J. Med. Plant* **2011**, *5*, 153–157. [CrossRef]
53. Hussein, M.M.; El-Gereadly, N.H.M. Infleunces of alpha tochopherol (vitamin E) and potassium hydrogen phosphate on growth and endogenous phytohormones of onion plants grown under salinity stress. *J. Agric. Sci.* **2007**, *32*, 9141–9151.
54. Venkatesan, K.; Selvakumari, P. Seasonal Influence of Seaweed Gel on Growth and Yield of Tomato (Solanum lycopersicum Mill.) Hybrid COTH 2. *Int. J. Curr. Microbiol. Appl. Sci.* **2017**, *6*, 55–66. [CrossRef]
55. Karjalainen, R.; Lehtinen, A.; Keinönen, M.; Julkunen-Tiitto, R.; Hietaniemi, V.; Pihlava, J.M.; Tiilikkala, K.; Jokinen, K. Benzothiadiazole and glycine betaine treatments enhance phenolic compound production in strawberry. *Acta Hortic.* **2002**, *567*, 353–356. [CrossRef]
56. Abdel-Mawgoud, A.M.R.; Tantaway, A.S.; Hafez, M.M.; Habib, H.A.M. Seaweed Extract Improves Growth, Yield and Quality of Different Watermelon Hybrids. *Res. J. Agric. Biol. Sci.* **2010**, *6*, 161–168.
57. Nguyen-Quoc, B.; Foyer, C.H. A role for "futile cycles" involving invertase and sucrose synthase in sucrose metabolism of tomato fruit. *J. Exp. Bot.* **2001**, *52*, 881–889. [CrossRef]
58. González, A.; Castro, J.; Vera, J.; Moenne, A. Seaweed Oligosaccharides Stimulate Plant Growth by Enhancing Carbon and Nitrogen Assimilation, Basal Metabolism, and Cell Division. *J. Plant Growth Regul.* **2013**, *32*, 443–448. [CrossRef]

© 2020 by the authors. Licensee MDPI, Basel, Switzerland. This article is an open access article distributed under the terms and conditions of the Creative Commons Attribution (CC BY) license (http://creativecommons.org/licenses/by/4.0/).

 *horticulturae*

Review

Impact of Grafting on Watermelon Fruit Maturity and Quality

Pinki Devi [1], Penelope Perkins-Veazie [2] and Carol Miles [1],*

[1] Northwestern Washington Research & Extension Center, Department of Horticulture,
     Washington State University, Mount Vernon, WA 98273, USA; pinki.devi@wsu.edu
[2] Department of Horticulture, Plants for Human Health Institute, North Carolina State University, Kannapolis,
     NC 28081, USA; penelope_perkins@ncsu.edu
*   Correspondence: milesc@wsu.edu

Received: 19 November 2020; Accepted: 4 December 2020; Published: 8 December 2020

Abstract: Watermelon (*Citrullus lanatus*) grafting has emerged as a promising biological management approach aimed at increasing tolerance to abiotic stressors, such as unfavorable environmental conditions. These conditions include environments that are too cold, wet, or dry, have soil nutrient deficiency or toxicity and soil or irrigation water salinity. Studies to date indicate that fruit yield and quality may be positively or negatively affected depending on rootstock-scion combination and growing environment. Growers need information regarding the general effect of rootstocks, as well as specific scion-rootstock interactions on fruit maturity and quality so they can select combinations best suited for their environment. This review summarizes the literature on watermelon grafting with a focus on abiotic stress tolerance and fruit maturity and quality with specific reference to hollow heart and hard seed formation, flesh firmness, total soluble solids, and lycopene content.

Keywords: flesh firmness; flowering; harvest time; lycopene; rootstock-scion combination; total soluble solids

---

1. Introduction

Grafting watermelon (*Citrullus lanatus*) onto disease resistant rootstock can minimize the problems associated with soil-borne diseases, such as Fusarium wilt (caused by *Fusarium oxysporum* f. sp. *niveum*) and Verticillium wilt (caused by *Verticillium dahliae*), as well as infection by root nematodes (*Meloidogyne incognita*) [1,2]. In addition, grafting can provide watermelon with tolerance against abiotic stressors, including temperature extremes, soil salinity, and nutrient deficiency or toxicity [3–6]. For example, grafting can be used to advance the planting date of watermelon due to tolerance of cool soil [1,7]. The root architecture of selected rootstocks enables better performance against abiotic stress factors through improved uptake of water and macro/micronutrients [5,8–11]. These recent advances in the understanding of rootstock-mediated effects on scion performance have broadened the application of grafting for the cultivation of watermelon under adverse environments.

To take advantage of grafting, growers need information regarding fruit maturity and quality (flesh firmness, total soluble solids (TSSs), lycopene content, etc.) in response to rootstock-scion combinations under various environmental conditions. For a non-climacteric fruit like watermelon, fruit maturity at harvest is especially important. The effect of rootstocks, scion-rootstock interaction, and environmental conditions on the timing of fruit maturity is needed to ensure fruit quality from grafted watermelon is optimized. Studies to date indicate that depending on rootstock-scion combination, fruit yield, and quality attributes may be either positively or negatively affected by grafting. However, most literature relies on the assumption of synchronous ripening in grafted and nongrafted plants, potentially leading to the confounding reports for fruit maturity and fruit quality resulting from grafting. Therefore, determining an appropriate harvest time for grafted watermelon is essential to maintain yield and quality stability.

2. Factors Affecting Time of Harvest

Harvest time may be affected when time of flowering is affected. Time of flowering may be impacted due to modification of hormonal signaling in response to the rootstock-scion combination of a grafted watermelon) [1,12–15]. Sakata et al. [14] reported that watermelon grafted onto bottle gourd (*Lagenaria siceraria*) has early formation of female flowers. In contrast, Davis et al. [1] reported a delay in flowering of up to 1 week with watermelon grafted on bottle gourd, resulting in an equal delay in fruit maturity. Delayed flowering was also observed for watermelon cv. Fujihikari, grafted onto pumpkin (*Cucurbita maxima*), wax gourd (*Benincasa hispida*), and interspecific hybrid squash rootstocks (*C. maxima × C. moschata*) compared to nongrafted treatments in the greenhouse [16]. Compared to nongrafted plants, female flowering of cv. Secretariat was delayed by 13 days and by 7 days when grafted onto bottle gourd cv. Round and interspecific hybrid squash cv., respectively, but there were no differences in days to harvest or fruit maturity [17]. The differences in flower initiation response could be due to hormonal signaling, combined with environmental effects that occur during graft union formation. For example, cytokinin is higher in grafted plants than nongrafted plants and could potentially have an effect on flowering time [11,18].

Because grafting may affect flowering and the harvest date, harvesting fruit from grafted and nongrafted plants concurrently may be the cause of conflicting results regarding fruit quality [1]. Most studies have assumed concurrent ripening of fruit from grafted and nongrafted plants, and fruit has been harvested at the same time for all treatments. Further, most studies reporting the effect of different rootstocks on fruit quality have not reported fruit maturity indices at harvest, such as senescence of leaflet and tendril attached to the fruit pedicel, skin color development, or ground spot formation. Yet for non-climacteric fruit, such as watermelon, fruit maturity at harvest has a major impact on fruit quality characteristics [18,19]. Kroggel and Kubota [20] reported that grafted fruit should be harvested 5 days after the leaflet and tendril are completely dry. Devi et al. [21] found that fruit quality of grafted watermelon was optimized with a 7-day delayed harvest from when those indicators of fruit physiological maturity occurred. The increase in time to harvest is likely due to the increased vigor of the rootstock, including increased water, oxygen, and mineral uptake [22,23]. Thus, reports of reduced fruit quality of grafted watermelon may not be an attribute imparted from grafting, rather reduced quality may be a factor of improper (early) harvest timing [24].

3. Effect of Grafting on Fruit Quality

Regardless of the advantages of grafting, there are still concerns that grafting could potentially reduce watermelon fruit quality below United States Department of Agriculture market standards of 10% soluble solid content (SSC) and 13 N flesh firmness [25]. Quality of grafted fruit must be equal to or better than that of nongrafted watermelon to ensure market acceptability [3,26,27]. Grafting can enhance fruit quality of watermelon by increasing the synthesis of endogenous hormones and the acquisition and transport of mineral nutrients [2,28] and also by altering secondary metabolites [2,29]. While fruit quality attributes from grafted plants have generally been found to be consistent with or slightly improved compared to nongrafted plants, results appear to be specific to rootstock-scion combinations under particular environmental conditions [30–35]. These effects contribute to contradictory reports of the impact of grafting on fruit quality [1]. Overall, most rootstocks, except bottle gourd, have been reported to increase watermelon fruit flesh firmness, total soluble solids (TSSs), and lycopene content compared to nongrafted fruit [26,36–39].

3.1. Hollow Heart and Hard Seed Formation

Hollow heart (placental detachment from the rest of the flesh) (Figure 1A) and hard seed formation (Figure 1B) are morphological abnormalities sometimes associated with grafted triploid watermelon and may reflect rootstock-scion incompatibility and/or adverse environmental conditions or cultural practices [1,16,22]. Seedless watermelon varieties are known to be more susceptible to hollow heart than seeded varieties, and currently, 95% of US watermelon production uses seedless cultivars [40,41]. Although there is no definitive cause, multiple factors, such as genetics, pollination, pollen viability, flowering time, decreased fruit tissue firmness, and environmental stressors, such as cold night temperatures, can cause this physiological disorder, especially in seedless watermelon [40,42,43]. A challenging problem with hollow heart disorder is that it is difficult to predict when it will occur. Johnson [40] reported that the primary cause of hollow heart is inadequate pollination. Triploid watermelon does not produce sufficient viable pollen and requires diploid (seeded) pollen sources [44]. When the distance from triploid plants to pollenizer plants increases, the incidence of hollow heart increases. Triploid plants located 1.5 m away from a pollenizer plant showed a hollow heart incidence of 56%, and plants located 2.4 m away from a pollenizer had a 74% incidence [40]. Prolonged periods of hot day-time temperatures also lower bee activity and pollination rates. Furthermore, pollen viability and rates of pollination are reduced by cold weather and prolonged periods of hot day-time temperature.

(**A**)   (**B**)

Figure 1. Hollow heart (**A**) and black seed formation (**B**) in grafted triploid watermelon.

Hollow heart is less likely to occur when fruit set is later in the season [43]. Grafting watermelon onto interspecific hybrid rootstock has been shown to increase fruit tissue firmness and may decrease the susceptibility of hollow heart formation in triploid watermelon [41]. For example, in North Carolina, where day/night temperature was 25 °C/15 °C, hollow heart incidence was reduced by 25–30% in fruit of triploid cv. Liberty grafted onto interspecific squash hybrid rootstock cvs. Carnivor and Kazako, compared to nongrafted fruit [45]. In contrast, Devi et al. [17] reported that in northwest Washington (16 °C/10.5 °C mean day/night temperature during the growing season), hollow heart formation was high (13–25 mm cracking, non-marketable) for cv. Secretariat nongrafted, as well as that grafted onto interspecific hybrid squash cvs. Tetsukabuto and Super Shintosa, bottle gourd cv. Pelop, and wax gourd cv. Round.

Immature white seed is commonly found in triploid watermelon, and under certain stress/environmental conditions, some seed coats may become hard and black and look like a mature seed but are empty inside [44,46] Unfavorable conditions, such as moisture extremes (drought, flood), fertilizer imbalance, or temperature extremes may cause the formation of hard seeds [44]. There is concern among growers that grafting may adversely increase the formation of hard seeds in triploid watermelon. Devi at al. [17] found six hard seeds per fruit on average for both grafted and nongrafted watermelon fruit in northwest Washington. At this study site, night temperature was 10.5 °C on average, which could account for the high incidence of hollow heart and formation of hard seed.

3.2. Flesh Firmness

Increased flesh firmness of grafted watermelon fruit has been shown in many studies [17,21,26,31,32,47–49], while no change was reported by some studies [50–52] Interspecific squash hybrid rootstocks appear to increase watermelon flesh firmness most consistently in both diploid and triploid scions [26,38,53]. Liu et al. [54] reported no difference in firmness when diploid watermelons were grafted onto five different *L. siceraria* and *Cucurbita ficifolia* rootstocks. In contrast, Yamasaki et al. [16] reported a significant increase in firmness when diploid watermelon cv. Fujihikari was grafted onto *C. maxima* × *C. moschata* hybrid rootstock than onto *L. siceraria*. Bruton et al. [26] reported increased firmness in both diploid and triploid watermelon when grafted onto *C. ficifolia* and *C. maxima* × *C. moschata* hybrid rootstocks, but *L. ciceraria* rootstock produced lower or more varied fruit firmness. Similarly, triploid watermelon cv. Secretariat grafted onto *C. maxima* × *C. moschata* hybrid, *L. siceraria*, and *Benincasa hispida* rootstocks had greater firmness than nongrafted fruit [17]. Increased flesh firmness in grafted watermelon could be due to higher cell density [55,56] and is a positive enhancement of fruit quality, as firm fruit have an extended shelf-life [26,57]. The impact of scion-rootstock genotypic effects on watermelon flesh firmness can be significant; therefore, choosing the right combination may be a necessary element for improving fruit quality and postharvest shelf-life [26,46,49,58]

3.3. Total Soluble Solids

Sugar and acids are the primary constituents of TSSs, in addition to small amounts of dissolved vitamins, fructans, proteins, pigments, phenolics, and minerals. TSSs are an important measure of sweetness of watermelon fruit [59]. While most research studies have generally shown that grafting has no effect on TSS content in watermelon fruit [7,17,21,31,60,61] the rootstock-scion combination can affect TSSs of grafted fruit [1,4,27]. For example, Petropoulos et al. [62] reported variable TSSs when watermelon cvs. Sugar Baby and Crimson Sweet were grafted onto *C. maxima* × *C. moschata* hybrid rootstocks. While most studies did not find a significant effect in TSS for grafted watermelon with *L. siceraria* rootstocks [33,35,36,63], 0.5–1.0% reduction in TSSs was reported for grafted compared to nongrafted watermelon treatments in several studies [26,32,35,47]. Only a few studies found grafting to have a negative impact on TSSs [64,65] and Davis and Perkins-Veazie [32] reported that TSSs were reduced only in their grafted diploid watermelon treatment.

3.4. Fruit pH and Titratable Acidity

Unlike SSC, which shows a continuous increase during watermelon fruit growth and development, fruit pH was at 4.6–4.7 until about 19 days postanthesis, which increased to 5.2 just before full red color and increased to 5.4 or greater at full ripeness [66]. The pH of fully ripe watermelon can differ with production area, ranging from 5.4 to 6.2, with a pH over 6.5 indicating overripe watermelon [67]. The pH of diploid watermelons grafted to an interspecific hybrid squash rootstock (cv. TZ148) was slightly lower than that of self-grafted fruit (5.7 and 5.8 pH, respectively) [24]. In a triploid watermelon (cv. Liberty), the pH of nongrafted or grafted fruit to an interspecific hybrid squash rootstock (cv. Carnivor) was 5.4 [45]. Titratable acidity was very low in watermelon, from 0.08–0.23% malic acid equivalents [24,68] In grafted watermelons, titratable acidity remained higher, even after ripeness, than in nongrafted fruit [24,39]. The slightly lower pH and higher titratable acidity of grafted fruit may be due to both the slight delay in fruit ripening and delayed fruit senescence, but neither had a significant effect on the marketability of fruit.

3.5. Lycopene Content

Watermelon is an important source of natural antioxidants due to lycopene, a fat-soluble carotenoid with valuable health-promoting benefits [69]. Lycopene content of red fleshed watermelon (4.81 mg/100 g on average) is almost 40% higher than that of tomato (3.03 mg/100 g) [69]. Lycopene content is an important measure of watermelon fruit quality and has been reported as higher [24,27,47,70], lower [32,37], and unchanged [17,21,26,31,32,52,65] in grafted fruit compared with nongrafted watermelon. Grafting was reported to increase lycopene content in diploid watermelon cv. Pegasus when grafted onto interspecific hybrid squash rootstock cv. TZ148 compared to nongrafted, where optimum harvest maturity was reached between 35 and 40 days after flower initiation in the nongrafted watermelon compared to 40–45 days for the grafted watermelon [24]. Further, a decrease in lycopene content as reported with *L. siceraria* and *C. argyrosperma* rootstocks [32,37]. Proietti et al. [27] reported a 40% increase in lycopene content for diploid mini-watermelon cv. Ingrid when grafted onto *C. maxima* × *C. moschata* hybrid cv. PS 1313 compared to nongrafted watermelon, whereas Bruton et al. [26] found only a 5% increase in lycopene content for triploid watermelon cv. Summer Sweet Brand 5244, grafted onto *C. maxima* × *C. moschata* cv. RS 1330 and *C. ficifolia* cv. RS 1420. These results demonstrate that lycopene content in grafted watermelon fruit is a rootstock-scion combination dependent trait. Differences in time to peak maturity of the cultivars in the different studies, combined with rootstocks from different genera, may influence lycopene content, possibly due to rootstock mediated regulation of lycopene metabolism [71]. Kong et al. [71] reported that *L. siceraria* rootstock enhanced lycopene content before the fully ripe stage in grafted watermelon by upregulating the lycopene biosynthetic genes, whereas wild watermelon (*C. lanatus* var. *citroides*) rootstock promoted lycopene accumulation in grafted watermelon fruit, potentially by downregulating lycopene catabolic genes at the fully ripe stage. Other research studies have shown that environmental conditions, such as soil fertility, irrigation, light intensity, and day/night temperatures, as well as harvest maturity and vine health, can affect lycopene formation, development, and stability in watermelon [69,72]. While numerous studies have evaluated lycopene content, the underlying mechanism of lycopene metabolism is still unclear in grafted watermelon fruit.

3.6. Citrulline

Although not present in most foods, the non-essential amino acid L-citrulline is found in high amounts in watermelon. Consumption of watermelon juice containing 4–6 g citrulline + arginine was found to effectively lower blood pressure and arterial stiffness [73,74]. The effects of grafting rootstock on subsequent fruit citrulline content was mixed. Soteriou et al. [23] reported higher amounts in grafted watermelon fruit (2.5 vs 3.0), although no significant differences in L-citrulline were found in 'Liberty' (2.7–3.1 in 2017 and 3.4 in 2019) [45]. The different results may be from different rootstock genetics and relative amounts of root-available potassium. Zhong et al. [75] found that nongrafted watermelon fruit were lower in citrulline than those grafted to wild watermelon, bottle gourd, or interspecific hybrid rootstocks when hydroponic solutions of 0.1- and 6-mM potassium were applied. The amount of citrulline increased at low levels of applied potassium, except for fruit grafted to the interspecific rootstock. Amounts of potassium found in the fruit tissue corresponded to application rates and were similar among grafted plants but were reduced for the low potassium application rate in nongrafted plants [76].

4. Conclusions

Research studies demonstrate that grafting can effectively mitigate the adverse effects of environmental stress on watermelon production. Abiotic stress tolerance of grafted plants is facilitated by the modifications in root traits, such as deeper and more extensive root systems, higher root hydraulic conductance, and faster induction of hormone accumulation (for example, abscisic acid). Rootstock/scion combinations can influence stress tolerance of grafted plants under different environmental conditions. Hormonal signaling plays an important role in graft union formation, rootstock–scion communication, growth and yield, and potential flowering and fruit quality. Fruit quality measures can be confounded by non-standardization of harvest maturity of grafted and nongrafted plants, and there is need for further research regarding timing of maturation to optimize fruit quality of grafted watermelon. Although the changes in fruit quality after grafting have been widely reported, the mechanisms involved in regulating fruit characteristics with different rootstocks are still unknown. Future research needs to focus on rootstock performance within specific growing regions, soil type, and weather conditions to enhance fruit quality and extend shelf-life. In addition, there is a need to evaluate multiple scion and rootstock combinations to better understand the interaction effects on fruit quality under varying environmental conditions. To fully understand the impact of grafting on fruit yield and quality, studies need to provide cultivar information and climate data. Furthermore, research efforts are needed to understand the diverse roles of post-grafting biological and metabolic processes at the molecular level and should take watermelon rootstock selection into consideration. Formation of hollow heart and hard seeds also need special attention in grafted triploid watermelon, as there is limited research that has addressed these two fruit quality issues. This information will enable growers to select scion and rootstock cultivars for consistency of performance across locations. Perhaps most importantly, there is a need to determine the impact of grafting on fruit development and ripening in order to recommend maturity indicators and timing for harvest to optimize fruit quality of grafted watermelon. Finally, researchers, agriculture extension specialists, and industry professionals need to work together to disseminate this information in order for growers to view grafting as an effective tool for producing high-quality watermelon under unfavorable environmental conditions.

Author Contributions: Conceptualization, C.M. and P.D.; writing—original draft preparation, P.D.; writing—review and editing, C.M., P.D., and P.P.-V. All authors have read and agreed to the published version of the manuscript.

Funding: Funding support was provided by the U.S. Department of Agriculture National Institute of Food and Agriculture (NIFA) Specialty Crop Research Initiative Grant 2016-51181-25404 and NIFA Hatch project 1017286.

Acknowledgments: We appreciate internal review by Lisa De Vetter, and technical support by Ed Scheenstra and Patti Kreider.

Conflicts of Interest: The authors declare no conflict of interest.

References

1. Davis, A.R.; Perkins-Veazie, P.; Sakata, Y.; Galarza, S.L.Ó.; Maroto, J.V.; Lee, S.G.; Huh, Y.C.; Miguel, A.; King, S.R.; Cohen, R.; et al. Cucurbit grafting. *Crit. Rev. Plant Sci.* **2008**, *27*, 50–74. [CrossRef]
2. Lee, J.M.; Kubota, C.; Tsao, S.J.; Bie, Z.; Echevarria, P.H.; Morra, L.; Oda, M. Current status of vegetable grafting: Diffusion, grafting techniques, automation. *Sci. Hortic.* **2010**, *127*, 93–105. [CrossRef]
3. Colla, G.; Rouphael, Y.; Leonardi, C.; Bie, Z. Role of grafting in vegetable crops grown under saline conditions. *Sci. Hortic.* **2010**, *127*, 147–155. [CrossRef]
4. Flores, F.B.; Sanchez-Bel, P.; Estan, M.T.; Martinez-Rodriguez, M.M.; Moyano, E.; Morales, B.; Bolarín, M.C. The effectiveness of grafting to improve tomato fruit quality. *Sci. Hortic.* **2010**, *125*, 211–217. [CrossRef]
5. Kumar, P.; Lucini, L.; Rouphael, Y.; Cardarelli, M.; Kalunke, R.M.; Colla, G. Insight into the role of grafting and arbuscular mycorrhiza on cadmium stress tolerance in tomato. *Front. Plant Sci.* **2015**, *6*, 477. [CrossRef] [PubMed]
6. Rouphael, Y.; Rea, E.; Cardarelli, M.; Bitterlich, M.; Schwarz, D.; Colla, G. Can adverse effects of acidity and aluminum toxicity be alleviated by appropriate rootstock selection in cucumber? *Front. Plant Sci.* **2016**, *7*, 1283. [CrossRef] [PubMed]
7. Miguel, A.; Maroto, J.V.; Bautista, A.S.; Baixauli, C.; Cebolla, V.; Pascual, B.; Lopez, S.; Guardiola, J.L. The grafting of triploid watermelon is an advantageous alternative to soil fumigation by methyl bromide for control of Fusarium wilt. *Sci. Hortic.* **2004**, *103*, 9–17. [CrossRef]
8. Cohen, R.; Dombrovsky, A.; Louws, F.J. Grafting as agrotechnology for reducing disease damage. In *Vegetable Grafting: Principles and Practices*; Colla, G., Pérez-Alfocea, F., Schwarz, D., Eds.; CAB International: Oxford, UK, 2017; pp. 155–170. [CrossRef]
9. Kumar, P.; Rouphael, Y.; Cardarelli, M.; Colla, G. Vegetable grafting as a tool to improve drought resistance and water use efficiency. *Front. Plant Sci.* **2017**, *8*, 1130. [CrossRef]
10. Louws, F.J.; Rivard, C.L.; Kubota, C. Grafting fruiting vegetables to manage soilborne pathogens, foliar pathogens, arthropods and weeds. *Sci. Hortic.* **2010**, *127*, 127–146. [CrossRef]
11. Nawaz, M.A.; Imtiaz, M.; Kong, Q.; Cheng, F.; Ahmed, W.; Huang, Y.; Bie, Z. Grafting: A technique to modify ion accumulation in horticultural crops. *Front. Plant Sci.* **2016**, *7*, 1457. [CrossRef]

12. Aloni, B.; Cohen, R.; Karni, L.; Aktas, H.; Edelstein, M. Hormonal signaling in rootstock-scion interactions. *Sci. Hortic.* **2010**, *127*, 119–126. [CrossRef]

13. Pulgar, G.; Villora, G.; Moreno, D.A.; Romero, L. Improving the mineral nutrition in grafted watermelon plants: Nitrogen metabolism. *Biol. Plant.* **2000**, *43*, 607–609. [CrossRef]

14. Sakata, Y.; Ohara, T.; Sugiyama, M. The history and present state of the grafting of cucurbitaceous vegetables in Japan. *Acta Hort.* **2007**, *731*, 159–170. [CrossRef]

15. Satoh, S. Inhibition of flowering of cucumber grafted on rooted squash stocks. *Physiol. Plant.* **1996**, *97*, 440–444. [CrossRef]

16. Yamasaki, A.; Yamashita, M.; Furuya, S. Mineral concentrations and cytokinin activity in the xylem exudate of grafted watermelons as affected by rootstocks and crop load. *J. Jpn. Soc. Hortic. Sci.* **1994**, *62*, 817–826. [CrossRef]

17. Devi, P.; Perkins-Veazie, P.; Miles, C.A. Rootstock and plastic mulch effect on watermelon flowering and fruit maturity in a *Verticillium dahliae* infested field. *HortScience* **2020**, *55*, 1438–1445. [CrossRef]

18. Kyriacou, M.C.; Rouphael, Y.; Colla, G.; Zrenner, R.; Schwarz, D. Vegetable grafting: The implications of a growing agronomic imperative for vegetable fruit quality and nutritive value. *Front. Plant Sci.* **2017**, *8*, 741. [CrossRef]

19. Kader, A.A. Flavor quality of fruits and vegetables. Perspective. *J. Sci. Food Agric.* **2008**, *88*, 1863–1868. [CrossRef]

20. Kroggel, M.; Kubota, C. *2017 Potential Yield Increase by Grafting for Watermelon Production in Arizona*; College of Agriculture, University of Arizona: Tucson, AZ, USA, 2017. Available online: https://repository.arizona.edu/bitstream/handle/10150/625432/az1732-2017_0.pdf?sequence=1&isAllowed=y (accessed on 20 April 2019).

21. Devi, P.; Lukas, S.; Miles, C. Fruit maturity and quality of splice-grafted and one-cotyledon grafted watermelon. *HortScience* **2020**, *55*, 1090–1098. [CrossRef]

22. Lee, J.M. Cultivation of grafted vegetables i. current status, grafting methods, and benefits. *HortScience* **1994**, *29*, 235–239. [CrossRef]

23. Masuda, M.; Gomi, K. Mineral absorption and oxygen consumption in grafted and non-grafted cucumbers. *J. Jpn. Soc. Hortic. Sci.* **1984**, *52*, 414–419. [CrossRef]

24. Soteriou, G.A.; Kyriacou, M.C.; Siomos, A.S.; Gerasopoulos, D. Evolution of watermelon fruit physicochemical and phytochemical composition during ripening as affected by grafting. *Food Chem.* **2014**, *165*, 282–289. [CrossRef] [PubMed]

25. U.S. Department of Agriculture. *United States Standards for Grades of Watermelons*; Agricultural Marketing Service, U.S. Department of Agriculture: Washington, DC, USA, 2006. Available online: https://www.ams.usda.gov/sites/default/files/media/Watermelon_Standard%5B1%5D.pdf (accessed on 3 April 2019).

26. Bruton, B.D.; Fish, W.W.; Roberts, W.; Popham, T.W. The influence of rootstock selection on fruit quality attributes of watermelon. *Open Food Sci. J.* **2009**, *3*, 15–34. [CrossRef]

27. Proietti, S.; Rouphael, Y.; Colla, G.; Cardarelli, M.; de Agazio, M.; Zacchini, M.; Rea, E.; Moscatello, S.; Battistelli, A. Fruit quality of mini-watermelon as affected by grafting and irrigation regimes. *J. Sci. Food Agric.* **2008**, *88*, 1107–1114. [CrossRef]

28. Wang, Q.; Men, L.; Gao, L.; Tian, Y. Effect of grafting and gypsum application on cucumber (*Cucumis sativus* L.) growth under saline water irrigation. *Agric. Water Manag.* **2017**, *188*, 79–90. [CrossRef]

29. Kakizaki, T.; Kitashiba, H.; Zou, Z.; Li, F.; Fukino, N.; Ohara, T.; Nishio, T.; Ishida, M. A 2-oxoglutarate-dependent dioxygenase mediates the biosynthesis of glucoraphasatin in radish. *Plant Physiol.* **2017**, *173*, 1583–1593. [CrossRef] [PubMed]

30. Cohen, R.; Burger, Y.; Horev, C. Introducing grafted cucurbits to modern agriculture: The Israeli experience. *Plant Dis.* **2007**, *91*, 916–923. [CrossRef]

31. Dabirian, S.; Inglis, D.; Miles, C. Grafting watermelon and using plastic mulch to control verticillium wilt caused by *Verticillium dahliae* in Washington. *HortScience* **2017**, *52*, 349–356. [CrossRef]

32. Davis, R.A.; Perkins-Veazie, P. Rootstock effects on plant vigor and watermelon fruit quality. *Cucurbit Genet. Coop. Rpt.* **2005**, *28–29*, 39–41. Available online: https://www.researchgate.net/publication/285707492_Rootstock_effects_on_plant_vigor_and_watermelon_fruit_quality (accessed on 15 July 2019).

33. Wimer, J.A.; Miles, C.A.; Inglis, D.A. Evaluating grafted watermelon for verticillium wilt severity, yield, and fruit quality in Washington State. *HortScience* **2015**, *50*, 1332–1337. [CrossRef]

34. Alexopoulos, A.A.; Kondylis, A.; Passam, H.C. Fruit yield and quality of watermelon in relation to grafting. *J. Food Agric. Environ.* **2007**, *5*, 178–179.

35. Rouphael, Y.; Schwarz, D.; Krumbein, A.; Colla, G. Impact of grafting on product quality of fruit vegetables. *Sci. Hortic.* **2010**, *127*, 172–179. [CrossRef]

36. Alan, O.; Zdemir, N.; Nen, Y. Effect of grafting on watermelon plant growth, yield and quality. *J. Agron.* **2007**, *6*, 362–365. [CrossRef]

37. Candir, E.; Yetişir, H.; Karaca, F.; Ustun, D. Phytochemical characteristics of grafted watermelon on different bottle gourds (*Lagenaria siceraria*) collected from the Mediterranean region of Turkey. *Turk. J. Agric.* **2013**, *37*, 443–456. [CrossRef]

38. Huitrón, M.V.; Ricárdez, M.; Dianez, F.; Camacho, F. Influence of grafted watermelon plant density on yield and quality in soil infested with melon necrotic spot virus. *HortScience* **2009**, *44*, 1838–1841. [CrossRef]

39. Kyriacou, M.C.; Soteriou, G.A.; Rouphael, Y.; Siomos, A.S.; Gerasopoulos, D. Configuration of watermelon fruit quality in response to rootstock-mediated harvest maturity and postharvest storage. *J. Sci. Food Agric.* **2016**, *96*, 2400–2409. [CrossRef]

40. Johnson, G. 2014 These Beautiful Watermelon Patterns are Driving Everyone Crazy. Available online: https://www.boredpanda.com/weird-watermelons-beautiful-hollow-heart/ (accessed on 22 August 2018).

41. Trandel, M.A.; Perkins-Veazie, P.; Schultheis, J.; Gunter, C.; Johannes, E. Grafting watermelon onto interspecific hybrid squash reduces hollow heart disorder. In *II International Symposium on Vegetable Grafting*; ISHS: Charlotte, NC, USA, 2019. Available online: https://www.researchgate.net/publication/336880681_Grafting_watermelon_to_interspecific_hybrid_rootstock_reduces_hollow_heart_disorder (accessed on 2 March 2018).

42. Kano, Y. Relationship between the occurrence of hollowing in watermelon and the size and the number of fruit cells and intercellular air spaces. *J. Jpn. Soc. Hortic. Sci.* **1993**, *62*, 103–112. [CrossRef]

43. Johnson, G. 2015 Research Finds Potential Cause of Hollow Heart Disorder in Watermelon. Available online: https://phys.org/news/2015-06-potential-hollow-heart-disorder-watermelons.html (accessed on 22 August 2018).

44. Maynard, D.N. 1992 Growing Seedless Watermelon Univ. FL. Coop. Ext. Serv. Fact Sheet HS687. Available online: http://edis.ifas.ufl.edu/CV006 (accessed on 2 December 2019).

45. Trandel, M.A. Cell Wall Polysaccharides in Grafted and Non-Grafted 'Liberty' Watermelon with Hollow Heart. Ph.D. Thesis, North Carolina State University, Raleigh, NC, USA, 2020. Available online: https://www.lib.ncsu.edu/resolver/1840.20/38167 (accessed on 12 September 2020).

46. Hodge, L. *Growing Seedless (Triploid) Watermelon*; Neb Guide; University of Nebraska–Lincoln Ext.: Lincoln, NE, USA, 1994; Available online: https://extensionpublications.unl.edu/assets/pdf/g1755.pdf (accessed on 20 April 2019).

47. Kyriacou, M.; Soteriou, G. Quality and postharvest performance of watermelon fruit in response to grafting on interspecific cucurbit rootstocks. *J. Food Qual.* **2015**, *38*, 21–29. [CrossRef]

48. Paroussi, G.; Bletsos, F.; Bardas, G.A.; Kouvelos, J.A.; Klonari, A. Control of fusarium and verticillium wilt of watermelon by grafting and its effect on fruit yield and quality. Proc. IIIrd Balkan Symp. Veg. and Potatoes. *Acta Hortic. (ISHS)* **2007**, *729*, 281–285. [CrossRef]

49. Yetisir, H.; Sari, N.; Y¨ucel, S. Rootstock resistance to fusarium wilt and effect on watermelon fruit yield and quality. *Phytoparasitica* **2003**, *31*, 163–169. [CrossRef]

50. Alan, O.; Sen, F.; Duzyaman, E. The effectiveness of growth cycles on improving fruit quality for grafted watermelon combinations. *Food Sci. Technol.* **2018**, *38*, 270–277. [CrossRef]

51. Buller, S.; Inglis, D.; Miles, C. Plant growth, fruit yield and quality, and tolerance to verticillium wilt of grafted watermelon and tomato in field production in the Pacific Northwest. *HortScience* **2013**, *48*, 1–7. [CrossRef]

52. Karaca, F.; Yetişir, H.; Solmaz, I.; Çandır, E.; Kurt, S.; Sari, N.; Guler, Z. Rootstock potential of Turkish *Lagenaria siceraria* germplasm for watermelon: Plant growth, yield and quality. *Turk. J. Agric. For.* **2012**, *36*, 167–177. Available online: https://journals.tubitak.gov.tr/agriculture/issues/tar-12-36-2/tar-36-2-4-1101-1716.pdf (accessed on 20 July 2018).

53. Soteriou, G.A.; Kyriacou, M.C. Rootstock-mediated effects on watermelon field performance and fruit quality characteristics. *Int. J. Veg. Sci.* **2015**, *21*, 344–362. [CrossRef]

54. Liu, R.Q.; Zhang, H.M.; Xu, J.H. Effects of rootstocks on growth and fruit quality of grafted watermelon. *J. Shanghai Jiaotong Univ. Agric. Sci.* **2003**, *21*, 289–294.

55. Fallik, E.; Ziv, C. How rootstock/scion combinations affect watermelon fruit quality after harvest? *J. Sci. Food Agric.* **2020**. [CrossRef]

56. Soteriou, G.A.; Siomos, A.S.; Gerasopoulos, D.; Rouphael, Y.; Georhiadou, S.; Kyriacou, M.C. Biochemical and histological contrition to texture changes in watermelon fruit modulated by grafting. *Food Chem.* **2017**, *237*, 133–140. [CrossRef]

57. Saftner, R.; Abbott, J.A.; Lester, G.; Vinyard, B. Sensory and analytical comparison of orange-fleshed honeydew to cantaloupe and green-fleshed honeydew for fresh-cut chunks. *Postharvest Biol. Technol.* **2006**, *42*, 150–160. [CrossRef]

58. Cushman, E.K.; Huan, J. Performance of four triploid watermelon cultivars grafted onto five rootstock genotypes: Yield and fruit quality under commercial growing conditions. *Acta Hortic.* **2008**, *782*, 335–342. [CrossRef]

59. Magwaza, L.S.; Opara, U.L. Analytical methods for determination of sugars and sweetness of horticultural products—A review. *Sci. Hortic.* **2015**, *184*, 179–192. [CrossRef]

60. Colla, G.; Rouphael, Y.; Cardarelli, M.; Massa, D.; Salerno, A.; Rea, E. Yield, fruit quality and mineral composition of grafted melon plants grown under saline conditions. *J. Hortic. Sci. Biotechnol.* **2006**, *81*, 146–152. [CrossRef]

61. Chouka, A.; Jebari, H. Effect of grafting on watermelon vegetative and root development, production and fruit quality. *Acta Hortic.* **1999**, *492*, 85–94. [CrossRef]

62. Petropoulos, S.A.; Khah, E.M.; Passam, H.C. Evaluation of rootstocks for watermelon grafting with reference to plant development, yield and fruit quality. *Int. J. Plant Prod.* **2012**, *6*, 481–492. [CrossRef]

63. Yetisir, H.; Sari, N. Effect of different rootstock on plant growth, yield and quality of watermelon. *Austral. J. Expt. Agric.* **2003**, *43*, 1269–1274. [CrossRef]

64. López-Galarza, S.; Batista, A.S.; Perez, D.M.; Miquel, A.; Baixauli, C.; Pascual, B.; Maroto, J.V.; Guardiola, J.L. Effects of grafting and cytokinin-induced fruit setting on colour and sugar-content traits in glasshouse-grown triploid watermelon. *J. Hortic. Sci. Biotechnol.* **2004**, *79*, 971–976. [CrossRef]

65. Turhan, A.; Ozman, N.; Kuscu, H.; Serbeci, M.S.; Seniz, V. Influence of rootstocks on yield and fruit characteristics and quality of watermelon. *Hortic. Environ. Biotechnol.* **2012**, *53*, 336–341. [CrossRef]

66. Corey, A.K.; Schlimme, D.V. Relationship of rind gloss and groundspot color to flesh quality of watermelon fruits during maturation. *Sci. Hortic.* **1988**, 211–218. [CrossRef]

67. Perkins-Veazie, P.; Collins, J.K.; Davis, A.R.; Roberts, W. Carotenoid content of 50 watermelon cultivars. *J. Agric. Food Chem.* **2006**, *54*, 2593–2597. [CrossRef]

68. Mao, L.; Jeong, J.; Que, F.; Huber, D.J. Physiological properties of fresh-cut watermelon (*Citrullus lanatus*) in response to 1-methylcyclopropene and post-processing calcium application. *J. Sci. Food. Agric.* **2006**, *86*, 46–53. [CrossRef]

69. Naz, A.; Butt, M.S.; Sultan, M.T.; Qayyum, M.M.; Niaz, R.S. Watermelon lycopene and allied health claims. *EXCLI J.* **2014**, *13*, 650–660. [CrossRef]

70. Perkins-Veazie, P.; Collins, J.K. Carotenoid changes of intact watermelon after storage. *J. Agric. Food Chem.* **2006**, *54*, 5868–5874. [CrossRef] [PubMed]

71. Kong, Q.; Yuan, J.; Gao, L.; Liu, P.; Cao, L.; Huang, Y.; Zhao, L.; Lv, H.; Bie, Z. Transcriptional regulation of lycopene metabolism mediated by rootstock during the ripening of grafted watermelons. *Food Chem.* **2017**, *214*, 406–411. [CrossRef] [PubMed]

72. Perkins-Veazie, P.; Collins, J.K.; Pair, S.D.; Roberts, W. Lycopene content differs among red-fleshed watermelon cultivars. *J. Sci. Food Agric.* **2001**, *81*, 983–987. [CrossRef]

73. Figueroa, A.; Sanchez-Gonzalez, M.A.; Perkins-Veazie, P.M.; Arjmandi, B.H. Effects of watermelon supplementation on aortic blood pressure and wave reflection in individuals with prehypertension: A pilot study. *Am. J. Hypertens.* **2011**, *24*, 40–44. [CrossRef]

74. Figueroa, A.; Sanchez-Gonzalez, M.A.; Wong, A.; Arjmandi, B.H. Watermelon extract supplementation reduces ankle blood pressure and carotid augmentation index in obese adults with prehypertension or hypertension. *Am. J. Hypertens.* **2012**, *25*, 640–643. [CrossRef]

75. Zhong, Y.; Shi, J.; Zheng, Z.; Nawaz, M.A.; Chen, C.; Cheng, F.; Kong, Q.; Bie, Z.; Huang, Y. NMR-based fruit metabonomic analysis of watermelon grafted onto different rootstocks under two potassium levels. *Sci. Hortic.* **2019**, *258*, 108793. [CrossRef]

76. Zhong, Y.; Chen, C.; Nawaz, M.A.; Jiao, Y.; Zheng, Z.; Shi, X.; Xie, W.; Yu, Y.; Guo, J.; Zhu, S.; et al. Using rootstock to increase watermelon fruit yield and quality at low potassium supply: A comprehensive analysis from agronomic, physiological and transcriptional perspective. *Sci. Hortic.* **2018**, *241*, 144–151. [CrossRef]

Publisher's Note: MDPI stays neutral with regard to jurisdictional claims in published maps and institutional affiliations.

© 2020 by the authors. Licensee MDPI, Basel, Switzerland. This article is an open access article distributed under the terms and conditions of the Creative Commons Attribution (CC BY) license (http://creativecommons.org/licenses/by/4.0/).

 *horticulturae*

Article

Correlations among Quality Characteristics of Green Asparagus Affected by the Application Methods of Elevated CO_2 Combined with MA Packaging

Li-Xia Wang [1,*], In-Lee Choi [1,2] and Ho-Min Kang [1,3,*]

[1] Division of Horticulture and Systems Engineering, Kangwon National University, Chuncheon 24341, Korea; cil1012@kangwon.ac.kr

[2] Agricultural and Life Science Research Institute, Kangwon National University, Chuncheon 24341, Korea

[3] Interdisciplinary Program in Smart Agriculture, Kangwon National University, Chuncheon 24341, Korea

[*] Correspondence: 2018lisa@kangwon.ac.kr (L.-X.W.); hominkang@kangwon.ac.kr (H.-M.K.);
Tel.: +82-33-250-6425 (H.-M.K.); Fax: +82-33-259-5562 (H.-M.K.)

Received: 9 November 2020; Accepted: 10 December 2020; Published: 14 December 2020

Abstract: This research investigated the effects of continuous elevated CO_2 (20%, (v/v)) application or a 3 day CO_2 pretreatment followed by modified atmosphere (MA) or micro-perforated (MP) packaging on the postharvest quality of asparagus. The combination of CO_2 pretreatment with MA packaging (Pre-MA) inhibited the yellowing of asparagus and fresh weight loss (FWL), whereas stem firmness slightly increased with all elevated CO_2 treatments. CO_2 pretreatments increased antioxidant activity in the stem, but not in the tip, in contrast to the continuous flow CO_2 (Flow-CO_2) treatment. The phenolic and flavonoid contents increased in the elevated CO_2 pretreatments and Flow-CO_2 treatment. The elevated CO_2 treatments, especially Flow-CO_2, inhibited the development of microorganisms, and the treated asparagus did not decay. Pre-MA and Flow-CO_2 treatments were more effective in maintaining the visual quality and retarding the off-odor of asparagus. Furthermore, significant correlations between sensory quality characteristics and physiological-biochemical attributes were recognized; three principal components were extracted and they explained 86.4% of asparagus characteristics. The results confirmed the importance of visual quality, off-odor, firmness, color parameters, SSC and total phenolic content. In conclusion, elevated CO_2 pretreatment followed by MA packaging (Pre-MA) was beneficial for extending asparagus cold storage shelf life, and Flow-CO_2 was the best treatment for inhibiting postharvest decay.

Keywords: elevated CO_2; modified atmosphere package; sensory and physiological-biochemical characteristics; total phenol; DPPH

1. Introduction

In recent years, the consumption of green asparagus (*Asparagus officinalis* L.) has been increasing due to its good eating quality, special flavor, and abundance of nutritional value. Unfortunately, rapid quality deterioration, including toughening, off-odor development, shriveling, and fresh weight loss due to higher respiration [1] and metabolic activities occurs. The rapid postharvest loss of green asparagus quality poses a challenge to the development of effective methods to retard the decline of quality and prolong shelf life. The application of modified atmosphere packaging (MAP) at low temperature as an effective technology has been reported to be beneficial for extending the shelf life and maintaining the quality of vegetables and fruits by reducing respiration rate and fresh weight loss, delaying ripening and minimizing physiological disorders and decays [2,3]. In lotus (*Nelumbo nucifera*), MAP treatments delayed the browning of roots, which involves changes in phenols [4]. The ambient gas levels in postharvest storage are important as they relate to respiration rate and physiological changes [5].

Prestorage or continuous elevated CO_2 treatments have been used for many fruits and vegetables to improve quality, inhibit the development of microbial groups and prolong shelf life. Short-term application of high CO_2 maintained the solids content and firmness of white asparagus spears, which is related to change of metabolic activity and respiration rate [6]. In grapes (*Vitis vinifera* L.), high CO_2 (20 kPa) pretreatment improved the appearance of bunches and maintained berry quality [7], reduced total decay and induced the accumulation of three small heat shock proteins (HSPs) [8]. Nutrients such as soluble sugar, soluble protein, and free amino acid content were also increased by elevated CO_2 in kidney beans (*Phaseolus vulgaris*) [9]. Treatment with 20% CO_2 prolonged the strawberry (*Fragaria X ananassa* Duch.) storage period to 12 days, reduced the energy charge related to the decline of $NADH/NAD^+$ and caused the accumulation of γ-aminobutyric acid (GABA) [10]. Fungal decay was prevented, and no fermentation was observed with 20% CO_2 pretreatment for two days of fresh goji (*Lycium* spp.) berries [11]. High CO_2 pretreatment has also been used for white asparagus, which confirmed that the effects of higher CO_2 (10%) on biochemical and textural properties were determined by the storage temperature [6]. In a previous study, we reported that vacuum packages supplemented with 60% (*v/v*) CO_2 induced a higher soluble solids content, lower firmness, greener color, and less weight loss, but more off-odor [12]. Considering the low risk and high benefit of high CO_2 treatment in many fruits and vegetables, the use of prestorage or continuous treatments with elevated CO_2 has been proposed as a suitable tool to control the development of microorganisms and maintain green asparagus quality. However, few studies have been carried out on the effect of elevated CO_2 treatment on green asparagus postharvest quality and changes in phytochemical component and antioxidant activity.

This work aimed to analyze the effectiveness of continuous 20% CO_2 treatment and CO_2 pretreatment followed by MAP in controlling the development of microbial groups and its effect on quality attributes, antioxidant ability, and physiological-biochemical characteristics during green asparagus storage period at 4 °C.

2. Materials and Methods

2.1. Plant Materials, CO_2 Treatments, Packaging Material and Storage Conditions

Green asparagus (*Asparagus officinalis* L. cv. 'Welcome') was cultivated and provided by a local farm (Yanggu-gun, Gangwon-do, Korea, lat. 38°12′33.00″ N, long. 127°13′3.00″ E) in May 2019. Green asparagus spears were harvested and transported to the Postharvest Physiology and Distribution Laboratory, Kangwon National University, on the same day and stored in a 4 °C refrigerator. Healthy and uniform (20–30 g/spear, 1.4 ± 0.1 cm diameter and 24.0 ± 1.0 cm length) green asparagus spears were selected and randomized for elevated CO_2 treatments in this study.

A passive modified atmosphere (MA) package with a 10,000 cc/m²·day⁻¹·atm⁻¹ oxygen transmission rate (OTR) (Dae Ryung Precision Packaging Industry Co., Ltd., Gwangju-si, Korea) and a micro-perforated (MP, 34 holes on 10,000 cc/m²·day⁻¹·atm⁻¹ MA films perforated by drilling with 0.6 mm diameter tips) package were selected to package asparagus spears. A 20% (*v/v*) carbon dioxide (CO_2, released from dry ice) treatment was applied in a sealed plastic box (30 × 21 × 15 cm) An air pump (DK-3000, Dae-Kwang Electronics Co., Ltd, Korea) was used to control the CO_2 concentration and CO_2 concentration was measured every 3 hours.

Initially, selected asparagus spears were randomly divided into two groups: pretreated with 20% CO_2 for 3 days and 20% CO_2 continuous application until the last day of cold storage (20 days). The 20% CO_2 pretreatment group was further divided into two groups: packaged with MA (Pre-MA) and MP (Pre-MP) packages. Spears treated with continuous 20% CO_2 (Flow-CO_2) were stored in a sealed plastic box (30–21–15 cm). Untreated groups packaged with MA (Cont-MA) and MP (Cont-MP) packages were used as the controls. All of the groups were stored at 4 ± 0.5 °C and 85 ± 5% relative humidity (RH) for 20 days. Five replicates of 20 green asparagus spears each were used for each packaging group.

2.2. Microbiology Analysis

Microbiological analysis was performed on initial samples (Initial day), after 3 days of pretreatment (Treated day 3), and after 20 days of storage according to Wang [12] with some modifications. Fresh asparagus (2.0 g) was mixed with 18 mL sterilized distilled water using a stomacher blender (Powermixer, B&F Korea, Gimpo-si, Korea) set at the highest speed (level 10, 200 rpm) for 3 min. Then, the mixture was diluted by a factor of 1000. Next, 1.0 mL of the dilution was dropped on a microbiology Petri film plate (3 M Co., St Paul, MN, USA). Aerobic bacteria, yeast and mold, and *Escherichia coli* were cultivated for 72 h at 35 °C, 72 h at 25 °C, and 24 h at 35 °C, respectively. The development of total aerobic bacteria (TAB), yeast and mold (Y&M), and *E. coli* were measured using Petrifilm Plate Reader (3M Co., St. Paul, MN, USA). The number of microorganisms was represented by the base 10-logarithm of the colony-forming unit concentration (log $CFU \cdot g^{-1}$).

2.3. Changes in Quality Parameters of Green Asparagus

Firmness (N) was measured at two locations on each asparagus spear, at 5 cm from the tip and 8 cm from the base using a rheometer (Compac-100, Sun Scientific Co. Led.,Tokyo, Japan) with a probe (Ø 3.0 mm) at 1.0 mm/sec speed.

Green asparagus color variables were measured using a color-difference meter (Model CR-400, Konica Minolta Sensing, Inc., Japan) at 5 cm from the tip and at 8 cm from the stem for lightness (L^*), Chroma (C^*), redness a^*, blueness b^* and hue angle (h°). The total color difference was represented as ΔE^*, and calculated as:

$$\Delta E* = \sqrt{(L*-L)^2 + (a*-a)^2 + (b*-b)^2} \tag{1}$$

where L, a and b were the color parameters of fresh asparagus spears without treatment (Initial day).

The total chlorophyll content was measured following Yoon [13] with slight modifications. Frozen asparagus samples (1.0 g) were chopped and mixed in 10 mL methanol and then incubated at 4 °C for 48 h to extract chlorophyll. The absorbances at 642.5 nm ($A_{642.5}$) and 660 nm (A_{660}) were measured using a UV–VIS spectrophotometer (UV mini model 1240, Shimadzu, Japan). The total chlorophyll content was calculated using the following formula:

$$\text{Total chlorophyll} \left(\text{mg} \cdot \text{mL}^{-1} \right) = 7.12 \times A_{660} + 16.8 \times A_{642.5} \tag{2}$$

Soluble solids content (SSC) was measured by a pocket refractometer (PAL-1, Atago, Tokyo, Japan). Asparagus samples were chopped up and extruded with gauze wrapping. The asparagus solution was directly dripped onto a pocket refractometer and the SSC result was indicated as °Brix at ambient temperature [13].

Fresh weight loss was measured according to the following formula:

$$\text{Weight loss}(\%) = \frac{\text{Initial fresh weight} - \text{Final fresh weight}}{\text{Initial fresh weight}} \times 100\% \tag{3}$$

Fresh asparagus was weighed and put in an oven (OF-21E, Jeio Tech Co., Ltd., Daejeon, Korea) for 10 min at 103 °C, dried to constant weight at 72 °C, and re-weighed. The water content of asparagus spears on the final day of storage at 4 °C was determined as follows:

$$\text{Water content}(\%) = \frac{\text{Fresh weight} - \text{Dry weight}}{\text{Fresh weight}} \times 100\%. \tag{4}$$

Sensory qualities, including visual quality and off-odor of green asparagus, were assessed by five skilled members from the "Postharvest Physiology and Distribution Laboratory" [13]. The visual quality of the asparagus was assessed throughout the entire storage period and scored on a scale of 1 to 5 (1 = worst: yellowing, decay, shrinking, woodiness, bract opening, 2 = bad, 3 = good, 4 = better,

5 = best: an at-harvest appearance, no decay or defects, dark green, no shrinking or bract opening). Off-odor was assessed after 20 days of storage and scored on a scale of 0 to 5 (0 = no off-odor and 5 = strong off-odor). Asparagus with a visual quality score equal to or greater than 3 and an off-odor score equal to or less than 3.0 was determined to be marketable.

2.4. The Effect of High CO_2 on Phenol, Flavonoids and DPPH of Green Asparagus

Asparagus (2.0 g) was homogenized in 1% HCl-methanol (*v/v*) solution, the homogenate was placed in a 25 mL tube with distilled water and incubated for 20 min in the dark, and then it was filtered with 110 mm filter paper. The absorbance of the solution was measured at 280 nm and 325 nm [14]. The contents of total phenol and flavonoids were calculated as follows:

$$\text{Total phenol content} = \frac{OD_{280}}{g} \tag{5}$$

$$\text{Flavonoids content} = \frac{OD_{325}}{g} \tag{6}$$

Antioxidant activity was measured by the DPPH (α,α-diphenyl-β-picrylhydrazyl) method [15]. Green asparagus samples (2.0 g) were ground at 4 °C in 20 mL of methanol and then filtered with 110 mm filter paper, 1.0 mL of the filtered solution was mixed with 1.0 mL of a 0.4 mM DPPH in methanol solution, and the mixture was incubated in darkness for 30 min before measuring the absorbance at 516 nm.

$$\text{DPPH radical scavenging activity (\%)} = \left[1 - \frac{\text{sample } A_{516}}{\text{blank } A_{516}} \right] \times 100\% \tag{7}$$

2.5. Gas Conditions and Respiration Rate

By inserting a needle into the packages through a septum, 1.0 mL gas samples from the headspace of packages were collected. Carbon dioxide (CO_2) concentration in different treatments was measured with an infrared CO_2/O_2 analyzer (Model Check Mate 9900, PBI-Dansensor, Ringsted, Denmark). Ethylene (C_2H_4) concentration was measured with a GC-2010 Shimadzu gas chromatograph (GC-2010, Shimadzu Corporation, Japan) [12], equipped with BP 20 Wax column (30 m × 0.25 mm × 0.25 um, SGE analytical science, Australia) and a flame ionization detector (FID). The detector and injector operated at 127 °C, the oven was at 50 °C, and the carrier gas (N_2) flow rate was 0.67 mL· s^{-1}. CO_2 and C_2H_4 concentration changes in MA and MP packages were determined on the first and third days of cold storage and subsequently were measured every 5 days up to 20 days of storage at 4 °C. The change in CO_2 concentration of pretreated, continuous and control groups was measured every 1 hour at ambient temperature and determined in five duplicate packages. The respiration rate was expressed as mL CO_2 kg^{-1} h^{-1}.

2.6. Statistical Data Analysis

The software Microsoft Excel 2019, R program (Version 4.0.2, 2020-06) and IBM SPSS Statistics (24, IBM Corp., Armonk, NY, USA) was used to statistically analyze the data. The effects of elevated CO_2, MA/MP packaging, and CO_2 × package interaction were analyzed using two-way ANOVA. All results are expressed as the means (n = 5) and their standard error (SE). Significant differences were tested with ANOVA (one-way analysis of variance) and Duncan's Multiple Range Test at $\alpha < 0.05$. All experiments were performed with at least three independent repetitions and repeated three times. Principal component analysis (PCA) scores and correlation coefficients between sensory quality and physiological and biochemical performance were analyzed using Rstudio version 4.0.2 (Team 2020), with distance measurements by Pearson's correlation coefficient test.

3. Results

3.1. Changes in Quality Parameters of Green Asparagus

3.1.1. Fresh Weight Loss (FWL)

Fresh weight loss developed in all treatments, with the increase greater in MP packages than in MA packages during the entire storage period (Figure 1). Fresh weight loss of green asparagus was 2.10 ± 0.06% in Cont-MP and 1.55 ± 0.09% in Pre-MP, whereas in MA it was 0.58 ± 0.05% in Cont-MA and 0.48 ± 0.02% in Pre-MA, all of which significantly differed. Flow-CO_2 treatment showed the highest FWL of 2.30 ± 0.05%, which was probably due to the lack of a film package during cold storage. It is worth noting that although the weight loss of different treatments was significantly different, all treatments had weight loss below 2.5% and were very similar to each other. FWL was well below the limit of marketable acceptance of 8%. In addition, elevated CO_2 inhibited FWL with a lower loss than that of the control. Significant differences were found between Cont-MP and Pre-MP. Pre-MA had the lowest FWL after 20 days of storage at 4 °C. The inhibition of FWL by high CO_2 pretreatments was also observed with goji berries, with significantly lower weight loss than that of control fruit [11]. Previous studies have reported that water content is important for modifying the quality and storage period of fruits and vegetables [16,17].

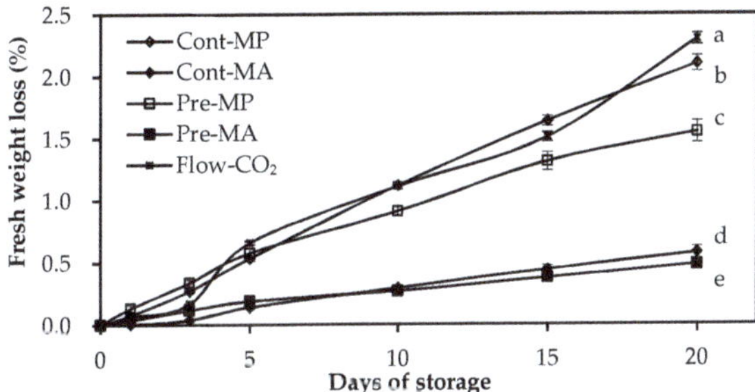

Figure 1. Fresh weight loss (%) during green asparagus cold storage as influenced by elevated CO_2 treatments. Values are means of five repetitions ± standard errors (SE). Different letters over bars indicate significant differences between treatments by Duncan's Multiple Range Test ($\alpha < 0.05$).

3.1.2. Firmness

Compared to the initial samples, asparagus firmness increased after 20 days of storage (Figure 2A). Considering the controls and all CO_2 treatments, a notable increase in firmness was observed in the stem with the initial value of 12.3 N. Compared to the control, Pre-MP and Flow-CO_2 treatments significantly increased the firmness on the last storage day and showed the highest firmness stem (~19.3 N). A significant difference in asparagus stem firmness was observed with elevated CO_2 pretreatment. The increase in firmness was less in MA than MP packages. This suggested that Pre-MA was more efficient at inhibiting the development of firmness during postharvest storage. This was in keeping with the results that the asparagus spear water content was lower in the Flow-CO_2 than the control and CO_2 pretreatments (data not shown). A similar negative relationship between firmness and water content in carrots has been reported [18]. The increase in firmness in asparagus was correlated with an increase in crude fibers, the development of lignin and the activity of phenylalanine ammonia-lyase [19]. Likewise, our results revealed significant differences in firmness only in the stem between the elevated CO_2 treatments and the control. This result suggests that CO_2 treatment influenced the wound response from cutting damage. Verlinden [20] reported a higher respiration rate, higher moisture loss and enhanced CO_2 diffusion during the wound response in the asparagus stem, which may lead to higher firmness.

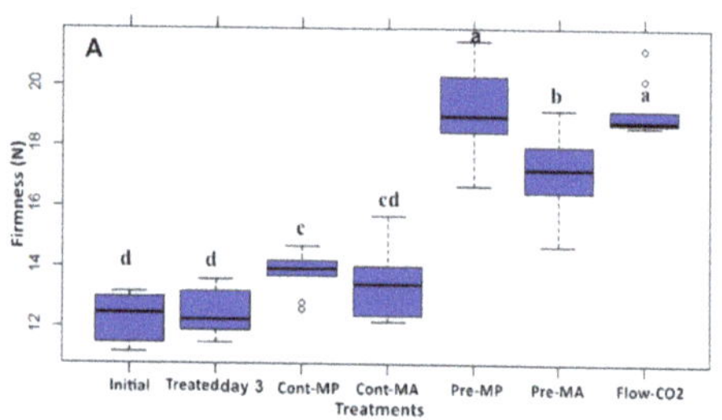

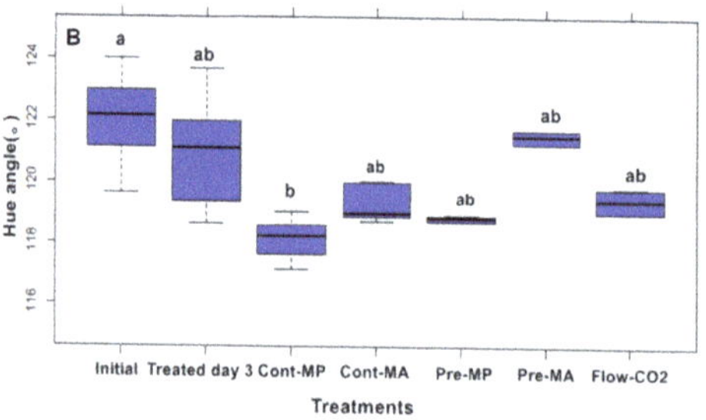

Figure 2. Effects of elevated CO_2 treatments on (**A**) firmness and (**B**) hue angle in asparagus stem on the initial day, treated day 3 and after 20 days of storage at 4 °C. The different letters represent significant differences by Duncan's Multiple Range Test ($\alpha < 0.05$ level). Box plots consist of the median, the lowest and the highest values, the outliers, and the approximate quartiles of our parameters.

3.1.3. Color Parameters

Color is one of the most important factors for postharvest biology and sensory evaluation of green asparagus. Yellowing (i.e., lower hue angle) of green asparagus indicates degradation of quality during storage [21]. Yellowing was obvious and showed a trend between the initial day and after 20 days of storage (Figure 2B) with only Cont-MP lower than the Initial value. The deviation from the raw material total color was represented as ΔE^* and the tip was significantly less than the other treatments at 20 days in the Flow-CO_2 treatment, and it did not differ from CO_2-treated day 3 (Table 1). For stems, the lowest color difference was observed for the Pre-MP treatment at 20 days, with a value of 5.19, which was less than the treated day 3 values.

Table 1. The means of chlorophyll content and total color deviation (ΔE^*) obtained from tip and stem of green asparagus in initial, treated day 3 and on the 20th day of storage at 4 °C.

Treatments	Chlorophyll Content (mg·mL^{-1})		ΔE^*	
	Tip	Stem	Tip	Stem
Initial day	5.66 a	2.18 a	-	-
Treated day 3	4.84 b	1.98 c	6.63 c	6.54 b
Cont -MP	3.23 f	1.12 g	9.35 b	10.23 a
Cont -MA	3.09 g	1.74 f	8.34 bc	10.30 a
Pre-MP	3.76 d	1.84 e	13.38 a	5.19 c
Pre-MA	4.68 c	2.11 b	7.29 bc	8.91 a
Flow-CO_2	3.62 e	1.89 d	6.25 c	8.20 ab

Values are means of five repetitions ($n = 5$). Different letters within a column indicate significant differences between treatments by Duncan's Multiple Range Test ($\alpha < 0.05$)

The green color is mainly determined by chlorophyll content [22]. The chlorophyll content was consistent with the hue angle value in all treatments (Table 1). Initial day samples had the highest chlorophyll content and subsequently exhibited a progressive decrease to the last day of cold storage. Interestingly, lower chlorophyll content was observed in the controls at the start and after cold storage. There was a significant difference among all the treatments; the Pre-MA treatment showed the highest chlorophyll content, and the Cont-MP treatment had the lowest content. The breakdown of chlorophyll is closely related to the remobilization of chloroplast proteins, lipids and metals during senescence [23]. An inhibitory effect of elevated CO_2 on the decrease in chlorophyll content of broccoli florets after harvest has been reported [24]. High CO_2 treatment affected anthocyanin stability and color expression in strawberries by changing the pH, while MA packaging with 15% CO_2 treatment maintained the anthocyanin stability of strawberries [25]. In the current study, Pre-MA treatment evidenced greater inhibition of yellowing during green asparagus cold storage. The mechanism of elevated CO_2 and MA package effects may have been partially based on a decrease in chlorophyllase activity that delays the degradation of chlorophyll because of the decrease in metabolic activity. This is consistent with the results that high CO_2 delayed chlorophyll degradation and anthocyanin accumulation by inhibiting chlorophyllase activity and downregulating the expression of *FaChl b reductase*, *FaPAO* and *FaRCCR*, which are related to chlorophyll state in postharvest storage of strawberry fruits [26]. Likewise, cold shock delayed the degradation of chlorophyll by suppressing the activity of chlorophyllase in cucumber [27].

3.1.4. Soluble Solids Content (SSC)

Total soluble solids content (SSC) are mainly sugars, which tended to decrease towards the end of shelf life in all of the samples [28]. A low soluble solids content was observed in all treatments and significantly differed among treatments (Table 2). On treated day 3, SSC had decreased to 5.58 °Brix compared to the initial day samples at 6.19 °Brix. The decrease was probably related to respiration, which is the key factor in converting soluble solids into energy [29]. There was no significant difference between the control (5.30 °Brix), Flow-CO_2 (5.23 °Brix) and Pre-MA (5.03 °Brix) treatments after 20 days of storage. However, the lowest SSC was in the Pre-MP treatment, with a content of 4.37 °Brix on the last day of cold storage. Flow-CO_2 treatment without film packages slowed the decrease in the SSC. Li and Zhang [30] reported that the water loss of green asparagus also contributed to an increasing soluble solids content, which is consistent with the water content of asparagus in this study (data not shown). A similar result was reported with grapes, in which high CO_2 treatment inhibited the decrease in the SSC due to an inhibition of normal postharvest metabolic activity [31].

Table 2. Effects of elevated CO_2 treatments on green asparagus soluble solids content (SSC), total phenolic and flavonoid contents, and DPPH-radical scavenging activity, on the initial day, treated day 3 and after 20 days of storage at 4 °C.

Treatments	SSC (°Brix)	Total Phenol (U·g^{-1})	Flavonoids (U·g^{-1})	DPPH (%)
Initial day	6.19 a	0.72 c	0.68 e	85.67 c
Treated day 3	5.58 b	0.82 a	0.91 a	92.67 a
Cont-MP	4.93 c	0.67 d	0.58 g	56.97 e
Cont-MA	5.30 bc	0.79 b	0.61 f	73.73 d
Pre-MP	4.37 d	0.80 ab	0.72 d	71.88 d
Pre-MA	5.03 c	0.81 a	0.74 c	88.64 b
Flow-CO_2	5.07 c	0.81 a	0.76 b	91.18 ab

Values are means of five repetitions (n = 5) and each letter in the same column indicate the significant differences by Duncan's Multiple Range Test ($\alpha < 0.05$).

3.2. The Effect of High CO_2 on Phenol, Flavonoids and DPPH of Green Asparagus

Phenolic compounds such as flavonoids and anthocyanin are important naturally occurring antioxidant compounds from many plants. The phenolic content could be influenced by biotic and abiotic factors [32,33]. Moreover, the antioxidant capacity of phenolic compounds is related to the ability to scavenge DPPH radicals in an assay [34]. In this study, total phenol and flavonoids content and DPPH radical scavenging activity were measured on the initial day, treated day 3 and after 20 days of storage (Table 2). The total phenolic and flavonoid content in the initial day samples were 0.72 U·g^{-1} and 0.68 U·g^{-1}, respectively. Total phenolic and flavonoid content increased in the CO_2 pretreatments on treated day 3, which showed the highest absorbance values of 0.82 U·g^{-1} and 0.91 U·g^{-1}, respectively. At 20 days, the levels of total phenolics under elevated CO_2 pretreatments had decreased compared to treated day 3 but increased compared to the initial day. The lowest total phenolic and flavonoid contents were observed in the Cont-MP treatment, with values of 0.67 U·g^{-1} and 0.58 U·g^{-1}, respectively. The difference between the control groups and all elevated CO_2 treatments was pronounced, although no difference was found between elevated CO_2 pretreatments and Flow-CO_2 treatment in total phenolics but not in flavonoids. Although elevated CO_2 treatments slowed the loss of phenolic components, a decrease occurred after longer storage in all elevated CO_2 treatments and control groups which is similar to changes in strawberry [35].

Postharvest CO_2 enrichment stimulated antioxidant activities and phenolic content in lettuce [36]. The beneficial effect may be because elevated CO_2 treatment is a kind of abiotic stress promoting the synthesis and accumulation of phenolic as a physiological response. Previously, CO_2 storage had marked effects on phenolic metabolites and quality, while modified atmospheres (MA) had a positive effect on phenolic-related quality [37]. High CO_2 may allow for the removal of free radicals, which are associated with an increase of antioxidant capacity [38]. The DPPH radical scavenging activity of green asparagus showed a similar change in pattern as the phenolic and flavonoid contents in response to the different treatments (Table 2). The highest DPPH value of 92.67 was a significant increase after CO_2 pretreatment. Flow-CO_2 treatment sustained the antioxidant activity through 20 days, with a DPPH value of 91.18. Based on these points, the Flow-CO_2 treatment was noted for improving asparagus qualities and maintaining the higher bioactive compound levels for long-term postharvest storage.

3.3. Gas Conditions in Packages and Respiration Rate

MA and MP packages involve the process of changing the atmospheric gas composition surrounding the green asparagus, especially carbon dioxide and oxygen concentration, by controlling gas exchange between the inside of the packages and the ambient air. The gas levels exhibited different concentration changes between MA packages and MP packages caused by their difference in oxygen transmission rates (OTRs). As shown in Figure 3A, the CO_2 concentration increased rapidly on the first day of cold storage in all packages and subsequently remained stable until a small decrease at 15 days of storage. MA packages showed higher CO_2 concentrations than MP packages throughout the whole storage period, 4–6% versus 0–2%, respectively. A CO_2 level above 10% and O_2 level below 3% can produce injuries to green asparagus at the optimal temperature [39,40]. Compared to the gas composition in MP packages, a better gas composition was observed in MA packages. The recommended gas concentration for green asparagus storage at 5 °C is 5–10% O_2, similar to ambient composition [41], and a CO_2 level of 4–6% probably decreases the respiration rate and inhibits anaerobic metabolism. The concentration of ethylene in all packages first increased (Figure 3B), and was followed by a decrease, finally becoming stable. It is possible that this was because elevated CO_2 reduced the respiration rate and inhibited the biosynthesis and negative effects of ethylene.

The respiration rate measured on the initial day was 26.26 mL kg^{-1} h^{-1} CO_2, while elevated CO_2 pretreatment showed a value of 24.62 mL kg^{-1}h^{-1} CO_2 (Figure 3C). Comparing the respiration rate of the Flow-CO_2 treatment on the final day (35.67 mL kg^{-1} h^{-1} CO_2) to the initial value, the significant difference could be ascribed to the long period of storage and the influence of elevated CO_2 since the respiration rate increased. Asparagus is classified as perishable due to a high respiration rate of 60 mg kg^{-1} h^{-1} CO_2 (at 5 °C) and rapid degradation of quality [5,42].

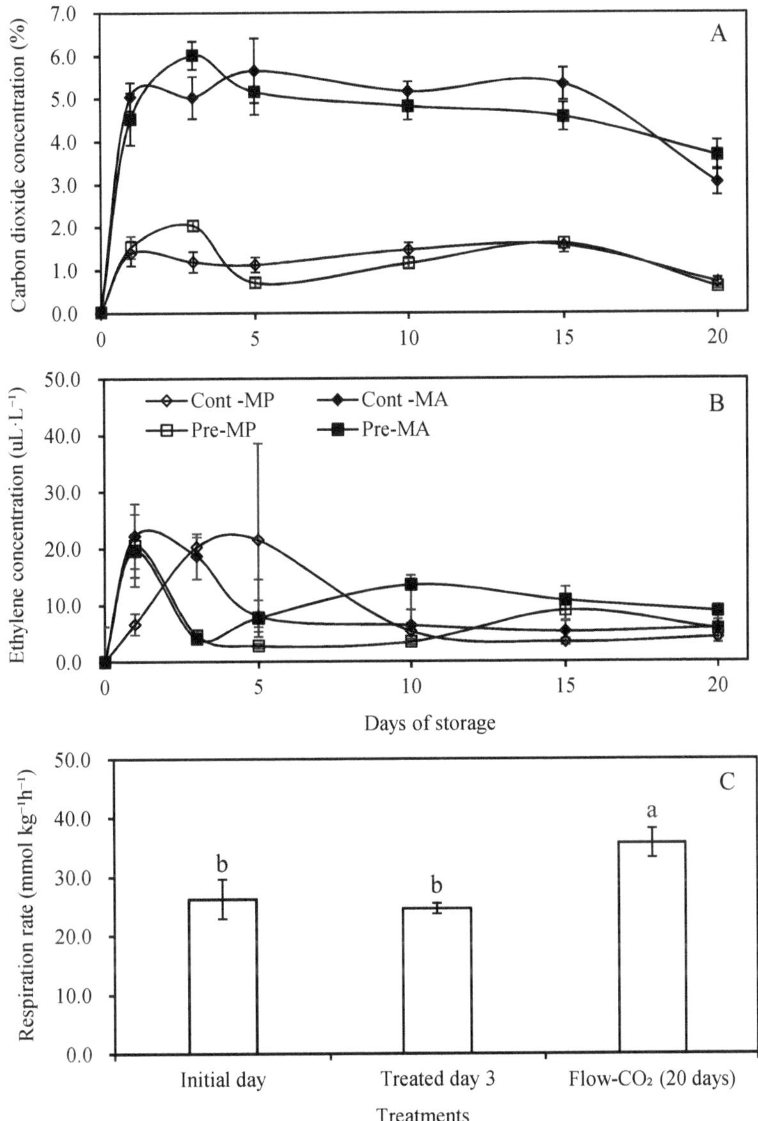

Figure 3. Changes in carbon dioxide (**A**) and ethylene (**B**) concentrations in modified atmosphere (MA) and micro-perforated (MP) packages during the postharvest storage, and the respiration rate (**C**) of green asparagus as influenced by high CO_2 pre and continuous treatments. Values are means of five repetitions ± standard errors (SE). Different letters over bars indicate significant differences among all treatments by Duncan's Multiple Range Test ($\alpha < 0.05$).

3.4. Microorganism Analysis

Compared to the initial value of 3.26 log $CFU \cdot g^{-1}$, the growth of total aerobic bacteria (TAB) was inhibited by elevated CO_2 pretreatment with 2.10 log $CFU \cdot g^{-1}$ (Table 3). Bacterial growth is critical in the spoilage of green asparagus, as significant differences in aerobic bacterial count between the initial and final storage days occurred. After 20 days of storage, TABs significantly increased, especially in Cont-MP, which had the highest count of 5.76 log $CFU \cdot g^{-1}$. Likewise, regardless of the packages, there was no significant difference in TABs between the CO_2 pretreatments and control groups. However, Flow-CO_2 with a TAB count of 2.16 log $CFU \cdot g^{-1}$ was not significantly different from the initial or treated day 3 samples, which indicated that the Flow-CO_2 treatment controlled the growth of these microorganisms and maintained lower counts up to the 20th day at 4 °C. The initial *E. coli* counts of green asparagus were 2.71 log $CFU \cdot g^{-1}$, and the counts increased through 20 days of storage in all treatments. Cont-MP and Cont-MA reached 4.31 and 4.20 log $CFU \cdot g^{-1}$, respectively. Treated day 3 did not show a sterilization effect on *E. coli* with 3.50 log $CFU \cdot g^{-1}$. However, a significant effect of elevated CO_2 pretreatments on *E. coli* was observed with lower counts after 20 days of storage at 4 °C than at treated day 3 (Table 3). Likewise, 3.01 log $CFU \cdot g^{-1}$ was observed under the Flow-CO_2 treatment on the last day of storage (20 days). The initial yeast and mold (Y&M) count before and after elevated CO_2 pretreatment was 0.00 log $CFU \cdot g^{-1}$ (Table 3). In contrast, Y&M increased significantly in the Cont-MP treatment on the last day of cold storage, with 2.10 log $CFU \cdot g^{-1}$. Notably, all elevated CO_2 treatments showed an encouraging lack of Y&M with counts of 0.00 log $CFU \cdot g^{-1}$. This was consistent with the result that high CO_2 was beneficial in controlling microbial growth in fresh-cut mango (*Mangifera indica*) cubes [43]. According to a previous study, short-term exposure to high levels of CO_2 modified the pathogenesis-related protein defense response and the low molecular mass chitinase, which renders grapes less susceptible to fungal infection [44]. It has also shown that elevated CO_2 retards the growth of microorganisms, which is lethal to spores and damages the cell wall of the microorganisms [45]. Likewise, the development of microorganisms inhibited by high CO_2 is probably due to its dissolution in the aqueous phase of food products, which causes intracellular acidification, inhibition of enzymatically catalyzed reactions and enzyme synthesis, and interaction with the cell membrane [46].

Table 3. Effects of elevated CO_2 pretreatments in MA and MP packages and continuous CO_2 treatments on the growth of microorganisms in green asparagus stored at 4 °C up to 20 days.

Treatments	Growth of Microorganisms (log $CFU \cdot g^{-1}$)		
	Total Aerobic Bacteria	*E. coli*	Yeast and Mold
Initial day	3.26 bc	2.71 c	0.00 b
Treated day 3	2.10 c	3.50 b	0.00 b
Cont-MP	5.76 a	4.31 a	2.10 a
Cont-MA	4.58 ab	4.20 a	0.00 b
Pre-MP	4.75 ab	3.34 b	0.00 b
Pre-MA	4.56 ab	3.45 b	0.00 b
Flow-CO_2	2.16 c	3.01 bc	0.00 b

Values are means of five repetitions (n = 5) and different letters in the same column indicate significant differences by Duncan's Multiple Range Test ($\alpha < 0.05$) between means.

3.5. Sensory Quality and Relative Impacts of Asparagus Quality Attributes

On the last storage day, asparagus quality characteristics were assessed. Figure 4 illustrates the relationships among important quality traits. Visual quality, off-odor, fresh weight loss, and stem firmness (F-Stem) were affected significantly by elevated CO_2 and different packages. In contrast, tip firmness (F-Tip), hue angle in the tip (h°-Tip) and stem (h°-Stem), soluble solids content (SSC), and water content were not affected by elevated CO_2 and different packaging treatments. Visual quality and off-odor are major factors that determine consumption. *p*-values in this study demonstrated that elevated CO_2 treatments and different packages had significant effects on nearly every quality characteristic (Table 4). The elevated CO_2 main effect, package main effect and $CO_2 \times$ package interaction significantly affected fresh weight loss of green asparagus whereas no effect ($p \geq 0.05$) was observed for water content. However, the elevated CO_2 main effect and interaction of $CO_2 \times$ package effects did affect tip hue angle (h°-Tip), and tip firmness (F-Tip).

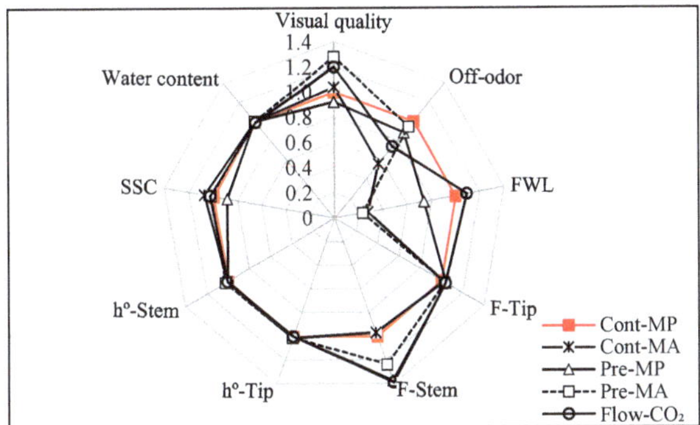

Figure 4. Spider graphs illustrating the relative impacts of elevated CO_2 treatments followed by MP or MA packaging on green asparagus quality characteristics after 20 days of storage at 4 °C. FWL: fresh weight loss; F-Tip: Tip firmness; F-Stem: Stem firmness; h°-Tip: Tip hue angle; h°- Stem: Stem hue angle; SSC: soluble solids content

Table 4. Analysis of variance results for the effects of elevated CO_2 × packaging (MP or MA) interaction green asparagus quality attributes after 20 days of storage at 4 °C.

Quality Attributes	CO_2	Package	CO_2 × Package
	p-Value	*p*-Value	*p*-Value
Visual quality	0.06	0.03 [z]*	0.07
Off-odor	0.04 *	0.03 *	0.00 **
Fresh weight loss	0.00 **	0.00 **	0.00 **
Water content	0.00 **	0.59	0.47
SSC	0.00 **	0.00 **	0.00 **
F-Tip [y]	0.32	0.77	0.41
F-Stem	0.00 **	0.03 *	0.16
h°-Tip	0.25	0.02 *	0.43
h°-Stem	0.00 **	0.00 **	0.00 **

[z]* = $p < 0.05$; ** = $p < 0.01$. [y] F-Tip: Tip firmness; F-Stem: Stem firmness; h°-Tip: Tip hue angle; h°-Stem: Stem hue angle.

Visual quality was estimated according to the extent of shrinkage, water loss, yellowing, bract opening, and fungal attack. In the present study, marketable visual quality was found in the Pre-MA and Flow-CO_2 treatments with scores of 3.2 and 3.0, respectively. Regarding the packages, the control groups lost retail value, and there was no significant difference between the control and Pre-MP treatments. The off-odor scores in the Cont-MA and Flow-CO_2 treatments were low and maintained commercial acceptability (off-odor score below 3.0), with scores of 1.9 and 2.5, respectively, after 20 days of storage. Regardless of packages, elevated CO_2 pretreatments lost marketable value with a strong off-odor. The accumulation of odor in green asparagus is caused by methanol and acetaldehyde, which are related to anaerobic respiration, microorganism decay and secondary metabolism [47,48]. Modified atmosphere (MA) packages reduced the production of off-odor by modifying the gas composition of packages and maintaining the adopted CO_2 (5–12%) content [5]. This is consistent with the growth of microorganisms in this study. Considering the better visual quality and slight off-odor, the notable efficiency of Flow-CO_2 treatment on asparagus quality and shelf life was confirmed in this study.

3.6. Correlations and Principal Components Analysis

The correlations between sensory quality characteristics and physiological and biochemical attributes after 20 days of storage at 4 °C are shown in Figure 5. A significant positive association between visual quality and soluble solids content (SSC), firmness (F), hue angle (h°), chlorophyll content (Chl), and DPPH in the tip (DPPH-Tip). Negative correlations between visual quality and off-odor, water content (WC), total phenolic content in stem (TP-Stem), total flavonoid content (Fla), and DPPH in the stem (DPPH-Stem) occurred. Likewise, the off-odor showed negative correlations with SSC, h°, Chl, and Fla. This was consistent with the performance of green asparagus in different treatments, i.e., the greener the color, the higher the soluble solids content and chlorophyll, the better the visual quality value and the less the off-odor. Another interesting negative correlation between firmness and WC and SSC was observed. A change in firmness is related to the change of cell structure caused by the transformation of SSC and storage sugars in cell walls [49]. The changes in cell structure affect the transpiration of plants which relates to water content and fresh weight loss [27]. In addition, negative correlations were observed between FWL and h° and Chl. Another important finding was the association between sensory quality and physiological-biochemical characteristics between tips and stems of asparagus, which could be related to the at-harvest wounding of the stem (the relations between the DPPH and TP).

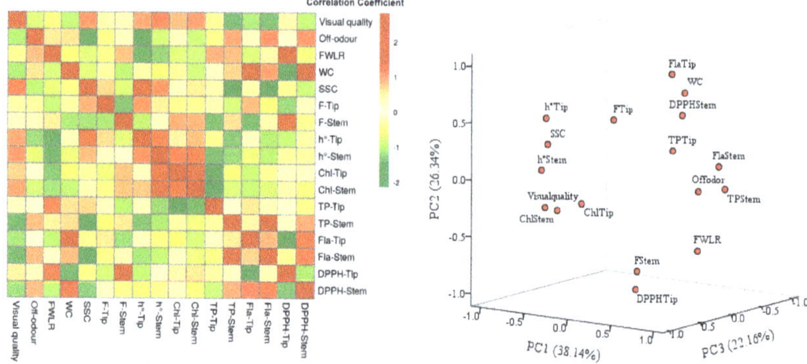

Figure 5. Pearson's correlation coefficient heatmap matrices and principal component analysis (PCA) among the responses of sensory, physiological quality attributes, and biochemical components in asparagus stored at 4 °C for 20 days of shelf life. FWL: fresh weight loss, WC: water content, SSC: soluble solids content, F-Tip: firmness in the tip, F-Stem: firmness in the stem, h°-Tip: Hue angle in the tip, h°-Stem: Hue angle in the stem, Chl-Tip: chlorophyll content in the tip, Chl-Tip: chlorophyll content in the stem, TP-tip: total phenolic content in the tip, TP-tip: total phenolic content in the stem, Fla-Tip: total flavonoid content in the tip, Fla-Tip: total flavonoid content in the stem, DPPH-Tip: DPPH radical scavenging activity in the tip, DPPH-Tip: DPPH radical scavenging activity in the stem.

The sensory quality of green asparagus was evaluated according to color, smell, decay, shrinkage, bract crack and others [50]. The association between sensory quality and the physiological situation was consistent with the sensory evaluation. These results are in agreement with Lu [51] who showed significant positive correlations between overall quality and firmness, SSC, TP, and negative correlations to the odor of 'Akihime' strawberry. Subsequently, principal component analysis (PCA) was performed and the results showed that the first three components (PC1, PC2, PC3) explained a total of 86.64% of the parameters of asparagus characteristic (Figure 5). The PC1, PC2, PC3 explained 38.14%, 26.34%, 22.16% of the variance, respectively. The visual quality, off-odor, TP-stem, Fla-Stem SSC, h° were clustered into the positive/negative of PC1, and the PC2 included FWL, DPPH, WC, Fla-Tip, F-Tip. Only F-Stem, TP-Tip, Chl-Tip variance belonged in PC3. The results confirmed the importance of visual quality, off-odor, firmness, color parameters, SSC and total phenolic content.

4. Conclusions

This study investigated the effects of elevated CO_2 pretreatment followed by modified atmosphere (Pre-MA) or micro-perforated (Pre-MP) packaging and continuous elevated CO_2 (Flow-CO_2) treatment on the quality characteristics of green asparagus during storage at 4 °C. The development of microbial groups was inhibited by elevated CO_2 treatments, especially the Flow-CO_2 treatment, which showed the lowest counts of microorganisms. Although higher firmness was observed, elevated CO_2 pretreatments and Flow-CO_2 treatment resulted in better sensory quality, strong inhibition of microorganisms, high content of the nutritional and antioxidant activity, a higher content of SSC, phenolics, chlorophyll, and DPPH, and lower respiration rates. Furthermore, the results from the heatmap coefficient analysis and principal component analysis indicated that sensory quality, physiological attributes, and antioxidant activity responded differently depending on the specific treatments. The elevated CO_2 treatments caused changes in physiological attributes and content of biochemical components that determined the sensory qualities of green asparagus. Overall, Flow-CO_2 or Pre-MA treatments had a significant effect on the qualities of green asparagus and could be useful for green asparagus cold storage.

Author Contributions: Conceptualization and methodology, L.-X.W., I.-L.C., and H.-M.K.; Experiments performance and data curation, L.-X.W.; writing, L.-X.W.; writing—review and editing, H.-M.K. All authors read and agreed to the published version of the manuscript.

Funding: This study was supported by the IPET through Export Promotion Technology Development Program, funding from the Ministry of Agriculture, Food and Rural Affairs (Nos 117035-03).

Acknowledgments: The authors would like to thank lab members for their assistance.

Conflicts of Interest: The author declares no conflict of interest.

References

1. Mangaraj, S.; Goswami, T.K. Modified atmosphere packaging of fruits and vegetables for extending shelf-life-A review. *Fresh Prod.* **2009**, *3*, 1–31.
2. Wang, J.W.; Jiang, Y.G.; Li, G.D.; Lv, M.; Zhou, X.; Zhou, Q.; Fu, W.W.; Zhang, L.; Chen, Y.F.; Ji, S.J. Effect of low temperature storage on energy and lipid metabolisms accompanying peel browning of 'Nanguo' pears during shelf life. *Postharvest Biol. Technol.* **2018**, *139*, 75–81. [CrossRef]
3. Yoon, H.S.; Choi, I.L.; Han, S.J.; Kim, J.Y.; Kang, H.M. Effects of Precooling and Packaging Methods on Quality of Asparagus Spears during Simulated Distribution. *Prot. Hortic. Plant Fact.* **2018**, *27*, 7–12. [CrossRef]
4. Liu, E.C.; Niu, L.F.; Yi, Y.; Wang, L.M.; Ai, Y.W.; Zhao, Y.; Wang, H.X.; Min, T. Expression Analysis of *ERFs* during Storage under Modified Atmosphere Packaging (High-concentration of CO_2) of Fresh-cut Lotus Root. *Hortscience* **2020**, *55*, 216–223. [CrossRef]
5. Kader, A.A. *Postharvest Technology of Horticultural Crops*, 3rd ed.; University of California, Agriculture and Natural Resources: Davis, CA, USA, 2002.
6. Huyskens-Keil, S.; Herppich, W.B. High CO_2 effects on postharvest biochemical and textural properties of white asparagus (*Asparagus officinalis* L.) spears. *Postharvest Biol. Technol.* **2013**, *75*, 45–53. [CrossRef]
7. Sanchez-Ballesta, M.T.; Romero, I.; Jimenez, J.B.; Orea, J.M.; Gonzalez-Urena, A.; Escribano, M.I.; Merodio, C. Involvement of the phenylpropanoid pathway in the response of table grapes to low temperature and high CO_2 levels. *Postharvest Biol. Technol.* **2007**, *46*, 29–35. [CrossRef]
8. Romero, I.; Casillas-Gonzalez, A.C.; Carrazana-Villalba, S.J.; Escribano, M.I.; Merodio, C.; Sanchez-Ballesta, M.T. Impact of high CO_2 levels on heat shock proteins during postharvest storage of table grapes at low temperature. Functional in vitro characterization of VVIHSP18.1. *Postharvest Biol. Technol.* **2018**, *145*, 108–116. [CrossRef]
9. Qian, L.; He, S.Q.; Liu, X.W.; Huang, Z.J.; Chen, F.J.; Gui, F.R. Effect of elevated CO_2 on the interaction between invasive thrips, Frankliniella occidentalis, and its host kidney bean, *Phaseolus Vulgaris*. *Pest Manag. Sci.* **2018**, *74*, 2773–2782. [CrossRef]
10. Li, D.; Li, L.; Xiao, G.N.; Limwachiranon, J.; Xu, Y.J.; Lu, H.Y.; Yang, D.M.; Luo, Z.S. Effects of elevated CO_2 on energy metabolism and gamma-aminobutyric acid shunt pathway in postharvest strawberry fruit. *Food Chem.* **2018**, *265*, 281–289. [CrossRef]
11. Kafkaletou, M.; Christopoulos, M.V.; Tsantili, E. Short-term treatments with high CO_2 and low O_2 concentrations on quality of fresh goji berries (Lycium barbarum L.) during cold storage. *J. Sci. Food Agric.* **2017**, *97*, 5194–5201. [CrossRef]

12. Wang, L.X.; Choi, I.L.; Lee, J.H.; Kang, H.M. The Effect of High CO_2 Treatment and MA Packaging on Asparagus Quality and Shelf Life during Cold Storage. *J. Agric. Life Environ. Sci.* **2019**, *31*, 41–49.

13. Yoon, H.S.; Choi, I.L.; Kang, H.M. Different Oxygen Transmission Rate Packing Films during Modified Atmosphere Storage: Effects on Asparagus Spear Quality. *Hortic. Sci. Technol.* **2017**, *35*, 314–322.

14. Cao, J.K.; Jiang, W.B.; Zhan, Y.M. *Experiment Guidance of Postharvest Physiology and Biochemistry of Fruits and Vegetables*, 1st ed.; China Light Industry Press: Beijing, China, 2007.

15. Oboh, G. Effect of blanching on the antioxidant properties of some tropical green leafy vegetables. *Lwt-Food Sci. Technol.* **2005**, *38*, 513–517. [CrossRef]

16. Sharma, H.K.; Kaur, J.; Sarkar, B.C.; Singh, C.; Singh, B.; Shitandi, A.A. Optimization of Pretreatment Conditions of Carrots to Maximize Juice Recovery by Response Surface Methodology. *J. Eng. Sci. Technol.* **2006**, *1*, 158–165.

17. Rashidi, M.; Bahri, M.H.; Abbassi, S. Effects of relative humidity, coating methods and storage periods on some qualitative characteristics of carrot during cold storage. *Am.-Eurasian J. Agric. Environ. Sci.* **2009**, *5*, 359–367.

18. Rashidi, M.; Gholami, M. Prediction of carrot firmness based on water content and total soluble solids of carrot. *Afr. J. Agric. Res.* **2011**, *6*, 1831–1834.

19. An, J.S.; Zhang, M.; Wang, S.J.; Tang, J.M. Physical, chemical and microbiological changes in stored green asparagus spears as affected by coating of silver nanoparticles-PVP. *Lwt-Food Sci. Technol.* **2008**, *41*, 1100–1107. [CrossRef]

20. Verlinden, S.; Silva, S.M.; Herner, R.C.; Beaudry, R.M. Time-dependent Changes in the Longitudinal Sugar and Respiratory Profiles of Asparagus Spears During Storage at 0 °C. *J. Am. Soc. Hortic. Sci.* **2014**, *139*, 339–348. [CrossRef]

21. Kaewsuksaeng, S.; Urano, Y.; Aiamlaor, S.; Shigyo, M.; Yamauchi, N. Effect of UV-B irradiation on chlorophyll-degrading enzyme activities and postharvest quality in stored lime (Citrus latifolia Tan.) fruit. *Postharvest Biol. Technol.* **2011**, *61*, 124–130. [CrossRef]

22. Zhao, S.S.; Yang, Z.; Zhang, L.; Luo, N.; Li, X. Effect of combined static magnetic field and cold water shock treatment on the physicochemical properties of cucumbers. *J. Food Eng.* **2018**, *217*, 24–33. [CrossRef]

23. Christ, B.; Hortensteiner, S. Mechanism and Significance of Chlorophyll Breakdown. *J. Plant Growth Regul.* **2014**, *33*, 4–20. [CrossRef]

24. Yamauchi, N.; Watada, A.E. Chlorophyll and xanthophyll changes in broccoli florets stored under elevated CO2 or ethylene-containing atmosphere. *HortScience* **1998**, *33*, 114–117. [CrossRef]

25. Nakata, Y.; Izumi, H. Microbiological and Quality Responses of Strawberry Fruit to High CO2 Controlled Atmosphere and Modified Atmosphere Storage. *Hortscience* **2020**, *55*, 386–391. [CrossRef]

26. Li, D.; Zhang, X.C.; Li, L.; Aghdam, M.S.; Wei, X.X.; Liu, J.Q.; Xu, Y.Q.; Luo, Z.S. Elevated CO2 delayed the chlorophyll degradation and anthocyanin accumulation in postharvest strawberry fruit. *Food Chem.* **2019**, *285*, 163–170. [CrossRef]

27. Zhao, S.S.; Yang, Z.; Zhang, L.; Luo, N.; Wang, C. Effects of different direction of temperature jump treatment on cucumbers. *J. Food Process Eng.* **2018**, *41*, e12600. [CrossRef]

28. Zurawicz, A.; Krzesinski, W.; Knaflewski, M. Changes in Soluble Solid Content in Green Asparagus Spears during Harvest Season. *Acta Hortic.* **2008**, *776*, 435–444. [CrossRef]

29. Nourian, F.; Ramaswamy, H.S.; Kushalappa, A.C. Kinetics of quality change associated with potatoes stored at different temperatures. *Lwt-Food Sci. Technol.* **2003**, *36*, 49–65. [CrossRef]

30. Li, T.H.; Zhang, M. Effects of modified atmosphere package (MAP) with a silicon gum film window on the quality of stored green asparagus (*Asparagus officinalis* L) spears. *Lwt-Food Sci. Technol.* **2015**, *60*, 1046–1053. [CrossRef]

31. Sanchez-Ballesta, M.T.; Jimenez, J.B.; Romero, I.; Orea, J.M.; Maldonado, R.; Urena, A.G.; Escribano, M.I.; Merodio, C. Effect of high CO2 pretreatment on quality, fungal decay and molecular regulation of stilbene phytoalexin biosynthesis in stored table grapes. *Postharvest Biol. Technol.* **2006**, *42*, 209–216. [CrossRef]

32. Mikulic-Petkovsek, M.; Slatnar, A.; Veberic, R.; Stampar, F.; Solar, A. Phenolic response in green walnut husk after the infection with bacteria Xanthomonas arboricola pv. juglandis. *Physiol. Mol. Plant Pathol.* **2011**, *76*, 159–165. [CrossRef]

33. Toivonen, P.M.A.; Hodges, D.M. Abiotic Stress in harvested fruits and vegetables. In *Abiotic Stress in Plants-Mechanisms and Adaptations*; Shanker, A., Venkateswarlu, B., Eds.; IntechOpen: London, UK, 2011; p. 442.

34. Kogure, K.; Goto, S.; Nishimura, M.; Yasumoto, M.; Abe, K.; Ohiwa, C.; Sassa, H.; Kusumi, T.; Terada, H. Mechanism of potent antiperoxidative effect of capsaicin. *Biochim. Biophys. Acta* **2002**, *1573*, 84–92. [CrossRef]

35. Chandra, D.; Lee, J.S.; Hong, Y.P.; Park, M.H.; Choi, A.J.; Kim, J.G. Short-term application of CO2 gas: Effects on physicochemical, microbial, and sensory qualities of "Charlotte" strawberry during storage. *J. Food Saf.* **2019**, *39*, e12597. [CrossRef]

36. Sgherri, C.; Pérez-López, U.; Micaelli, F.; Miranda-Apodaca, J.; Mena-Petite, A.; Muñoz-Rueda, A.; Quartacci, M.F. Elevated CO_2 and salinity are responsible for phenolics-enrichment in two differently pigmented lettuces. *Plant Physiol. Biochem.* **2017**, *115*, 269–278. [CrossRef] [PubMed]

37. Tomás-Barberán, F.A.; Espín, J.C. Phenolic compounds and related enzymes as determinants of quality in fruits and vegetables. *J. Sci. Food Agric.* **2001**, *81*, 853–876. [CrossRef]

38. Wang, S.Y.; Bunce, J.A.; Maas, J.L. Elevated carbon dioxide increases contents of antioxidant compounds in field-grown strawberries. *J. Agric. Food Chem.* **2003**, *51*, 4315–4320. [CrossRef]

39. Gariepy, Y.; Raghavan, G.S.V.; Castaigne, F.; Arul, J.; WillemotI, C. Precooling and Modified Atmosphere Storage of Green Asparagus. *J. Food Process. Preserv.* **1991**, *15*, 215–224. [CrossRef]

40. Simón, A.; Gonzalez-Fandos, E. Influence of modified atmosphere packaging and storage temperature on the sensory and microbiological quality of fresh peeled white asparagus. *Food Control* **2011**, *22*, 369–374. [CrossRef]

41. Bartz, J.A.; Brecht, J.K. *Postharvest Physiology and Pathology of Vegetables*; CRC Press: Boca Raton, FL, USA, 2002.

42. Lipton, W.J. Postharvest Biology of Fresh Asparagus. In *Horticultural Reviews*; Janick, J., Ed.; JohnWiley and Sons: Hoboken, NJ, USA, 2011; pp. 69–155.

43. Jutatip, P.; Hidemi, I. Shelf Life and Microbial Quality of Fresh-cut Mango Cubes Stored in High CO_2 Atmospheres. *J. Food Sci.* **2005**, *70*, 6.

44. Vazquez-Hernandez, M.; Navarro, S.; Sanchez-Ballesta, M.T.; Merodio, C.; Escribano, M.I. Short-term high CO_2 treatment reduces water loss and decay by modulating defense proteins and organic osmolytes in Cardinal table grape after cold storage and shelf-life. *Sci. Hortic.* **2018**, *234*, 27–35. [CrossRef]

45. Rao, L.; Bi, X.F.; Zhao, F.; Wu, J.H.; Hu, X.S.; Liao, X.J. Effect of high-pressure CO_2 processing on bacterial spores. *Crit. Rev. Food Sci. Nutr.* **2016**, *56*, 1808–1825. [CrossRef]

46. Mohn, G. Modified atmospheres. In *The Microbiological Safety and Quality of Foods*; Lund, B.M., Baird-Parker, A.C., Gould, G.W., Eds.; Springer Science & Business Media: Gaithersburg, MD, USA; Aspen, CO, USA, 2000; pp. 214–234.

47. Tudela, J.A.; Marin, A.; Garrido, Y.; Cantwell, M.; Medina-Martinez, M.S.; Gil, M.I. Off-odor development in modified atmosphere packaged baby spinach is an unresolved problem. *Postharvest Biol. Technol.* **2013**, *75*, 75–85. [CrossRef]

48. Lin, Q.; Lu, Y.Y.; Zhang, J.; Liu, W.; Guan, W.Q.; Wang, Z.D. Effects of high CO2 in-package treatment on flavor, quality and antioxidant activity of button mushroom (Agaricus bisporus) during postharvest storage. *Postharvest Biol. Technol.* **2017**, *123*, 112–118. [CrossRef]

49. Herppich, W.B.; Huyskens-Keil, S. Cell wall biochemistry and biomechanics of harvested white asparagus shoots as affected by temperature. *Ann. Appl. Biol.* **2008**, *152*, 377–388. [CrossRef]

50. Herppich, W.B.; Huyskens-Keil, S.; Hassenberg, K. Impact of Ethanol Treatment on the Chemical Properties of Cell Walls and Their Influence on Toughness of White Asparagus (Asparagus officinalis L.) Spears. *Food Bioprocess Technol.* **2015**, *8*, 1476–1484. [CrossRef]

51. Lu, H.Y.; Wang, K.D.; Wang, L.; Li, D.; Yan, J.W.; Ban, Z.J.; Luo, Z.S.; Li, L.; Yang, D.M. Effect of superatmospheric oxygen exposure on strawberry (Fragaria x ananassa Fuch.) volatiles, sensory and chemical attributes. *Postharvest Biol. Technol.* **2018**, *142*, 60–71. [CrossRef]

Publisher's Note: MDPI stays neutral with regard to jurisdictional claims in published maps and institutional affiliations.

 © 2020 by the authors. Licensee MDPI, Basel, Switzerland. This article is an open access article distributed under the terms and conditions of the Creative Commons Attribution (CC BY) license (http://creativecommons.org/licenses/by/4.0/).

MDPI

St. Alban-Anlage 66

4052 Basel

Switzerland

Tel. +41 61 683 77 34

Fax +41 61 302 89 18

www.mdpi.com

Horticulturae Editorial Office

E-mail: horticulturae@mdpi.com

www.mdpi.com/journal/horticulturae

www.ingramcontent.com/pod-product-compliance
Lightning Source LLC
La Vergne TN
LVHW070849170726
843515LV00005B/608